TRULY PECULIAR

The Economist Explains

FANTASTIC FACTS
that are
STRANGER THAN FICTION

Edited by
TOM STANDAGE

Published in 2021 under exclusive licence from The Economist by
Profile Books Ltd
29 Cloth Fair
London EC1A 7JQ
www.profilebooks.com

Typeset in Milo by MacGuru Ltd

Printed and bound in Great Britain by CPI Group (UK) Ltd, Croydon CRO 4YY

A CIP catalogue record for this book is available from the British Library

ISBN 978 1 78816 896 0
eISBN 978 1 78283 893 7

Contents

Introduction: in search of the truly peculiar 1

Truly peculiar: fantastic facts that are stranger than fiction 3

How football can reduce civil wars 5

Are overweight politicians less trustworthy? 7

How Nigeria's love of fancy wigs fuels a global trade in hair 10

How the pandemic has changed illegal-drug habits 12

Why it's better to be a poor pupil in a rich country than the reverse 14

How Big Macs can measure the true size of China's economy 16

What bank-robbery looks like in the digital age 19

Why Zoom meetings are so dissatisfying 21

Quantifying the size of the Bank of Mum and Dad 23

What bosses can learn from a hostage negotiator 26

Globally curious: particular propensities from around the world 29

Why Corsican number plates are so popular in France 31

Why tea sales have boomed in Johannesburg's restaurants 33

Why self-help books are so popular in China 35

Why the Baltic states are reconfiguring their electric grids 37

Why ant-egg soup is a nostalgic treat in Laos 39

Why is control of Western Sahara so controversial? 41
Where are the world's most expensive cities? 44
What did Lava Jato, Brazil's anti-corruption investigation, achieve? 46
Why western European armies have shrunk dramatically 49
Where does Britain's royal family get its money from? 52
Why Spain's high-speed trains are such poor value 54

The art of the possible: politics, local and global 57
Why is Washington, DC not a state? 59
Why has civil war returned to Ethiopia? 62
In America, far-right terrorist plots outnumbered far-left ones in 2020 64
How are maritime boundaries determined? 67
A rift in democratic attitudes is opening up around the world 70
Why Taiwan is not recognised on the international stage 73
Faith in government declines when mobile-internet access arrives 75

Questions of faith: religion and belief 77
Why Muslim scholars disagree over keeping dogs as pets 79
How personal freedom varies across the Islamic world 81
Why so many Muslim leaders are building grand mosques 83
Why American politicians are more pious than their constituents 85
What do the French mean by laïcité? 88
Why there are more Christians in China than in France or Germany 90
Which countries still have blasphemy laws? 93

Gender agenda: sex, marriage and equality 95
The sad truth about Vegas weddings 97
Why it's so hard to predict the peak of world population 99
What is being done to tackle "period poverty" in the West? 102

Why east and west German women still work vastly different hours 104
Why the lot of female executives is improving 107
How to shrink America's gender pay gap 109
Why women are less likely than men to die from covid-19 111
Why America's demographics are looking more European 114

Medically speaking: health, death and disease 117
Americans are driving less, so why are more of them dying in accidents? 119
Why are rich countries so vulnerable to covid-19? 121
How different kinds of vaccines work 123
Why the world desperately needs more nurses 126
Why heatwaves are killing a record number of people 129
Why is turkey meat becoming more popular? 131
How the Japanese boosted their longevity by balancing their diets 133
Why playing video games in lockdown is good for your mental health 136

Follow the money: economical, with the truth 139
Why China's economic centre of gravity is moving south 141
Why Black Friday isn't what it used to be 143
What is the African Continental Free Trade Area? 145
Why money buys happiness, but euphoria comes dear 147
What is the fuss over central-bank digital currencies? 149
Why people from poor countries pay more for visas 152
When bribery pays – and when it doesn't 154

Speaking my language: words and wisdom 157
Why Japanese names flipped (for Western readers) in 2020 159
Where did the word "robot" come from? 161
Why 2020 was Twitter users' most miserable year yet 164
Why Joe Biden's folksy speaking style is a strength, not a weakness 166

Do languages evolve in the same way as biological species? 168
Is "irregardless" really a word? 171
What wine vocabulary reveals about the nature of language 174
Why Wikipedia's future lies in poorer countries 177
Why is the definition of genocide controversial? 180
Why is America's best journalism published at the end of the year? 182

Facts of the matter: science, nature and the environment 185
Why Lithuanians suddenly embraced recycling 187
Why it matters that the Laptev Sea is freezing later 189
Why green investing has so little impact 191
Why more people are making it to the top of Mount Everest than ever 193
Why the age of forests, not just their area, is important 195
Why dung beetles' love of human faeces results in scientific bias 198
Why the Nobel prize delay is growing 200
How the shark forgot his skeleton 202
How to spot dodgy academic journals 204

Playing by the rules: sport, games and leisure 207
How empty stadiums made it possible to measure referees' bias 209
Why African countries issue stamps celebrating English cricketers 211
Why are penalties becoming more common in elite football? 213
How new swing techniques are revolutionising golf 215
Why are so many athletics records falling? 218
Why engineers, not racers, are the true drivers of success in motor sport 220
Why young chess stars always usurp the old 223

Media matters: arts and culture 225
Why dead rappers now get a bigger sales boost 227
Why musicians are increasingly collaborating with their fans 229
How the cola wars became a cultural phenomenon 232
How Hollywood is losing ground in China 234
What accounts for the strange appeal of snow globes? 236
How does Christmas television vary around the world? 239
What is the world's most frequently aired television programme? 242

Contributors 244
Index 245

Introduction: in search of the truly peculiar

HERE IS A PECULIAR FACT: the title of this book was suggested by an algorithm. It sounds like something out of a science-fiction story or an avant-garde novel, but it is true. And it seems appropriate for this book, which rounds up fantastic facts and extraordinary explanations that are true, yet are peculiar enough to be stranger than fiction. For example, how can football reduce civil wars? Why are Corsican number plates suddenly so popular in France? And how can Big Macs measure the true size of China's economy? And, for that matter, how can a book be named by an algorithm?

Previous books in this series, all of which present collections of explainers and charts from *The Economist*, have been called *Seriously Curious*, *Uncommon Knowledge* and *Unconventional Wisdom*. Each title is a play on words that combines a word that means "facts", "factual" or "factually" with a word that means "unusual" or "unusually". There are three distinct combinations: "unusual facts", "unusually factual" and "factually unusual". The three preceding titles correspond to one of these combinations: "Seriously Curious" means something similar to "factually unusual", while "Uncommon Knowledge" and "Unconventional Wisdom" are both equivalent to "unusual facts". Once this underlying formula has been identified, writing a program to generate more titles in a similar vein is a simple matter.

I wrote the program in Python, a popular language that helpfully has a powerful add-on library called Natural Language Toolkit, or NLTK. One of NLTK's features is the ability to generate a list of synonyms of a particular word. Starting with a list of synonyms for "unusual", my program expanded it by adding all the synonyms of those synonyms. It then did the same for "unusually", "facts", "factual" and "factually". It then generated random book titles by picking synonyms from these lists to produce variations on the formulae "unusual facts", "unusually factual" and "factually unusual".

Many of the resulting suggestions, it must be said, were not suitable book titles, possibly because the synonyms-of-synonyms approach had the effect of casting the net rather wider than anticipated. That is why you are not holding a book called "Disadvantageously Drunk", "Oddly Elucidate", "Unknown Brainstorm", "Extraneous Sapience", "Unfeignedly Foreign" or "Amazingly Honourable". But among the avalanche of awful suggestions were some good ideas: "Oddly Informative", "Strangely Instructive", "Surprisingly Enlightening", "Unusually Illuminating", "Truly Unusual", "Remarkably True" – and "Truly Peculiar". I sent a shortlist of the most promising algorithmic suggestions to the publisher, and *Truly Peculiar* was the title we chose.

As you can see, the story of how an algorithm came to name this book is slightly technical, but can be explained in a way that is easily grasped by a non-specialist. It highlights something going on in the world today, namely that computers are becoming more capable of manipulating language. Understanding it can give you insight into a field or a topic that you might not be familiar with – and may well cause you to raise your eyebrows. It is, in other words, a good example of the kind of truly peculiar explanations you will find more of inside this book. We hope you enjoy reading them.

Tom Standage
Deputy Editor, *The Economist*
April 2021

Truly peculiar: fantastic facts that are stranger than fiction

How football can reduce civil wars

Could a missed penalty kick really have helped bring peace to Ivory Coast? In 2005 its national football team was on the brink of qualifying for the World Cup for the first time. Having won its final qualifying match, Ivory Coast just needed Cameroon to lose or draw the match it was playing against Egypt. The awarding of a late penalty set the Cameroonians up for a win. But Pierre Womé hit the post. The ball flew wide. Ivory Coast was in.

Listening to the match on the radio, the Ivorian players erupted in cheers. Then they pleaded for peace in their war-torn country. "We proved today that all Ivorians can coexist and play together," said Didier Drogba, the captain. The team knelt. "We beg you on our knees... please lay down your weapons and hold elections," said Mr Drogba. The clip was played again and again on Ivorian television. In the months that followed, the warring parties began talking and, eventually, agreed to a ceasefire. In 2007 they agreed to peace.

There were, of course, factors in play other than Ivory Coast's win, Mr Womé's missed shot and Mr Drogba's impassioned plea. But, according to a study published in *American Economic Review* in 2020, the outcomes of important football matches can have a dramatic effect on national unity and, thus, civil wars.

The study's authors, led by Emilio Depetris-Chauvin of the Pontifical Catholic University of Chile, looked at how Africans identified themselves and how much they said they trusted each other in the days after important national-team matches. They found that people surveyed after their national squad had won were 37% less likely to identify primarily with their ethnic group, and 30% more likely to trust other ethnicities, than those interviewed just before. "This is entirely driven by national-team victories, whereas defeats have no discernible impact on that self-identification," the authors noted.

The bigger the match, the bigger the boost to national solidarity and trust. This does not merely reflect a general post-victory euphoria, according to the study. Incumbent politicians and ruling parties got no bounce in approval from a win. Nor was there any impact on respondents' optimism about the economy.

Victories also lead to a reduction in violence. The authors compared countries that narrowly qualified for the African Cup of Nations in recent years with those that narrowly missed out. The countries that squeaked in experienced almost 10% less conflict in the next six months than those that did not. The make-up of the squad probably matters, too. Mr Drogba noted that his team hailed "from the north, south, centre and west" of Ivory Coast. "The effect of victories is stronger the more diverse the ethnic composition of the national team," the study's authors wrote.

So could more football reduce conflict in Africa? Perhaps. But the positive results hold only for high-stakes matches, not friendlies (matches unrelated to a competition). And the bonhomie can be fleeting. A second civil war broke out in Ivory Coast in 2010 – though calm returned in 2011, after Mr Drogba and many others once again appealed for peace.

Are overweight politicians less trustworthy?

In the southern English town of High Wycombe, the local member of parliament, mayor and councillors are weighed publicly every year in the town centre, in a self-described attempt to deter them from "gaining weight at taxpayers' expense". Cheers erupt if a lawmaker has shed a few pounds; boos await those who have put any on. The centuries-old tradition is conducted largely in jest. But the townsfolk might be on to something.

The more overweight the government, the more corrupt the country, according to a study of 15 post-Soviet states published in July 2020. The study's author, Pavlo Blavatskyy of Montpellier Business School in France, used an algorithm to analyse photographs of almost 300 cabinet ministers and estimate their body-mass index (BMI), a measure of obesity. He found that the median BMI of a country's cabinet was highly correlated with its level of corruption, as measured by indices compiled by the World Bank and Transparency International.

According to these measures, the least corrupt post-Soviet countries were the three Baltic states – Estonia, Lithuania and Latvia – and Georgia. These four countries also boasted the slimmest cabinets. Meanwhile, Tajikistan, Turkmenistan and Uzbekistan ranked worst for corruption. The latter two countries also had the chubbiest ministers, according to Mr Blavatskyy's estimates, along with Ukraine. Obesity is surprisingly common among the upper ranks of government in the former Soviet Union. Nearly one-third of ministers across all 15 states were rated "severely obese" by Mr Blavatskyy's algorithm, including 54% of Uzbek and 44% of Tajik ministers. Only 3% were estimated to be in the normal weight category. Despite Vladimir Putin's carefully cultivated reputation as a powerful, healthy strongman, his cabinet was just as flabby as those in neighbouring countries. Perhaps, like Shakespeare's Julius Caesar, he prefers to have about him men that are fat, fearing the "hungry" look of the lean men. He is wrong, Mr Blavatskyy's data suggest. The tubby may look content, but appear to be more grasping.

Centres of gravity

Post-Soviet countries

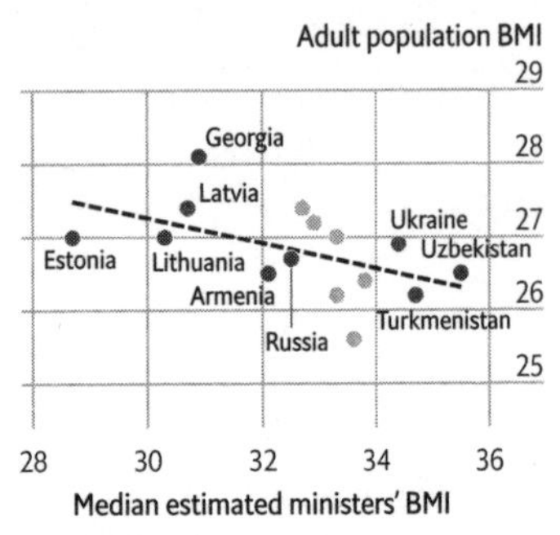

Source: "Obesity of politicians and corruption in post-Soviet countries", P. Blavatskyy, *Economics of Transition and Institutional Change*, 2020

Corruption is notoriously difficult to quantify. Most measures rely on the perceptions of the public, which tend to change slowly. Thus, a proxy variable – like ministers' BMIs – can prove useful, especially in the short term. Mr Blavatskyy notes that despite undergoing a revolution that set it on the path to democracy in 2018, Armenia's score in Transparency International's corruption index remained unchanged from the previous year. And yet the estimated BMI of ministers in the new Armenian government was slightly lower than its pre-revolution value – a sign, in this analysis, that the country may have begun to root out corruption.

Might not fleshier politicians simply come from places with heavier people? In fact, the opposite is true: countries with more-obese ministers tend to have less-obese people (see the right-hand chart). Indeed, the post-Soviet countries deemed most corrupt by Transparency International also have the poorest and most underweight populations.

Of course, correlation is not causation. Obese politicians are not necessarily more dishonest than their slimmer counterparts. But it might be wise for officials in the former eastern bloc to slim down all the same. Studies show that voters are less likely to plump for candidates who are, well, plump. Even if most politicians manage to avoid having to step on the scales before their subjects, like those in High Wycombe, the link between health and electability should weigh on their minds nonetheless.

How Nigeria's love of fancy wigs fuels a global trade in hair

"My outfit for the day determines what hair I will be wearing," says Olayinka Titilope, a Nigerian wigmaker. She has a different peruke for each day of the month. The weather also influences her choice. On cooler days she might opt for long, thick locks. During the summer she tends towards lighter bob-cuts. Ms Titilope hopes her hairdos will inspire the customers who visit her wig gallery in downtown Lagos, Nigeria's commercial capital. She sells wigs for $60-800. The most expensive ones are made of human hair from Cambodia, she says.

Some African feminists argue that to wear a long, straight-haired wig or hair extension is to grovel to Western ideals of beauty. Yet wig-buyers in Nigeria seem to enjoy variety. Sellers advertise hair from everywhere. Brazilian hair is praised for its sheen and durability; Vietnamese, for its bounce; Mongolian, because it is easy to curl. One seller in Lagos offers "Italian posh hair", which is supposedly odour-free. Whatever the label says, much of the hair really comes from elsewhere – often China, a source some buyers deem downmarket. It is hard even for the most conscientious hair-traders to determine or trace where their wares came from. Most of the hair that reaches Africa travels via factories in China, where it is sorted and often treated, dyed or curled. Bundles of human hair may be bulked up with horse mane or goat thatch. Chinese locks are sometimes packed into boxes labelled "Peruvian hair", which is coveted in Nigeria. Responsible shopkeepers must pick a good supplier and hope for the best. Those with fewer scruples rebrand the hair once it arrives.

Demand in Nigeria is huge, but not everyone wants to pay Nigerian tariffs. Benin, a popular route for goods smuggled into its much bigger neighbour, sucked in 11% of the world's fake-hair imports in 2018 – some 50 times what might be expected, given its tiny population. Nigeria itself shipped in more than 3,600 tonnes of hair (including human, animal and synthetic hair, as well as ready-made hair pieces). If even half of that was from human scalps, it would amount to the waist-length locks of more than 10m people. "The demand for hair generally exceeds supply, fuelling an almost constant sense of scarcity," writes Emma Tarlo in her book *Entanglement: The Secret Lives of Hair*.

In the past decade hair exports from Myanmar have quadrupled in volume, making it the world's fourth-largest exporter. Nay Lin, a hair-trader in the former capital, Yangon, says he knows when the economy is bad because more women turn up at his shop to sell their tresses. "Today I have had ten heads so far," he says, a lot for one day. A pile of dark bunches glistens on the floor beside him. Clients earn around $18 for their hair, though prices vary according to weight. Most of it gets shipped to China, but he is unsure where it goes after that. Mr Lin exported duck feathers until he discovered that hair was more lucrative. Meanwhile, some 500km north of Yangon, in the town of Pyawbwe, farmers who once harvested onions and chillies now spend their days unpicking hairballs. These are often gathered by door-to-door collectors, who buy hair from people's combs and bathroom plugs. Some hairballs arrive in sacks from India and Bangladesh. Workers in Pyawbwe (which has earned the nickname "Hair City") make about $1.20 a day untangling them and removing lice or white strands. This hair is so common in Chinese factories that it is referred to as "standard hair". It costs more than the fake stuff, but less than locks cut straight from a head.

"We call that stuff factory trash," scoffs Ms Titilope, who insists that none of it goes into her products. She does not like using Indian hair, either, because much of it is shorn off pilgrims and some customers think it is cursed. Most Hindus will have their heads shaved at a temple at least once as a symbol of surrendering their egos to Vishnu, a god. The temples then sell the tresses. But in Nigeria some believe that snakes may have slithered over the hair. To Ms Titilope's customers, hair is an important status symbol. Women with silky locks tumbling down their backs stroll past others with coarse, synthetic threads. On the streets of Lagos, wigs reflect wealth.

How the pandemic has changed illegal-drug habits

When Seth Rogen, an actor, was asked in April 2020 how he had been passing the time in lockdown, he replied, true to character, "I've smoked a truly ungodly amount of weed." A report published in September 2020 suggests he was not alone. According to the Global Drug Survey, an annual study of global drug-use trends, as covid-19 left millions stuck indoors, anxious and bored, many people turned to psychoactive substances.

Mr Rogen's drug of choice saw the biggest rise in usage. According to the online survey, which polled some 55,000 people in 11 mostly rich countries, almost two in five respondents had increased their use of cannabis since the start of the pandemic. Australia saw the greatest increase, at 49%, with America (46%) and Britain (44%) close behind. Respondents largely ascribed their greater use of the drug to the extra free time and boredom of lockdowns. But pre-existing mental-health conditions also played a role: among those with mental illnesses who upped their toking, 41% cited stress and 38% cited depression (among cannabis users without those conditions, the figures were 20% and 15%, respectively). Unsurprisingly, respondents in countries hit hard by the pandemic, such as America, were more likely to say they got stoned more often because they were depressed.

Party drugs, meanwhile, were found to have fallen out of favour, in large part because of the abrupt disappearance of nightclubs and festivals, as well as the difficulty of meeting dealers during lockdown. The survey found that ecstasy usage was down by 41%, while cocaine was down by 38% and ketamine by 34%. Few seemed to miss their habits. Nearly a third of cocaine users said their mental health was better as a result of taking less of the stuff; a quarter of ecstasy enthusiasts said the same. Around two-fifths of those who used one or other drug said they didn't feel like taking them during a pandemic.

Whether these changes in drug habits will outlast the pandemic

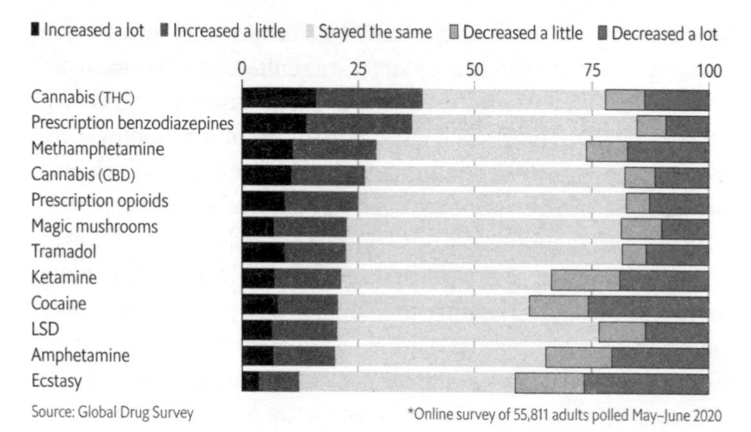

remains to be seen, but illegal party drugs are likely to bounce back when nightclubs reopen and festivals resume. In the meantime, experts have warned against self-medicating with cannabis. Sharing drugs can help spread the virus, and heavy smoking can increase the risk of complications from covid-19. Perhaps Mr Rogen should avoid passing his spliffs around.

Why it's better to be a poor pupil in a rich country than the reverse

There are some things money can't buy. Education, however, does not appear to be among them – at least as measured by performance on international exams. On average, pupils in wealthy countries obtain vastly higher test scores than those in developing ones. Strong students tend, in turn, to become productive workers, making the mostly rich economies they join richer still.

The exact mechanism by which knowledge can be bought remains unclear. Do students in the rich world fare better because their governments provide superior schools? Or is the reason that they tend to have richer parents, and enjoy more educational resources at home? A working paper by Dev Patel of Harvard University and Justin Sandefur of the Centre for Global Development, a think-tank, published in September 2020, offers strong evidence that the wealth of a country affects exam results just as much as the wealth of an individual pupil's household does.

Evaluating test scores around the world is harder than it sounds. Although pupils in the rich world mostly take one of a few big international exams, many developing countries rely on regional tests. This precludes apples-to-apples comparisons.

The authors surmounted this obstacle by setting an exam in 2016 for 2,314 children in Bihar, in northern India. It included a mixture of questions from the leading tests and ones taken from smaller exams. Using answers from the same pupils on the same day to questions from different tests, they built a statistical model they called a "Rosetta Stone". It can translate scores from a range of exams – such as one used solely in west Africa – into an equivalent mark on other common international tests.

The researchers then used their model to estimate how pupils in 80 different countries would fare on the benchmark Trends in International Mathematics and Science Study. Their data show that the wealth of a student's country and family have similar impacts on test scores – but that means, in turn, that big gaps between countries in GDP per person matter more than smaller ones in household income. For example, Russian students whose families earn $2,000 a year (measured in 2005 dollars), putting them at the lower end of the income scale in that country, score better than students from families in Morocco, a poorer country, whose families earn $4,000 a year. The overall wealth of the country, in short, has more of an impact on educational outcomes than the relative wealth of the family.

That said, the study also found that the influence of parental earnings is not constant. An extra $1,000 in family income "buys" a larger increase in test scores in highly unequal countries than it does in ones that split their economic pie more evenly. One possible reason is that elites tend to educate their children privately in places where wealth is concentrated, such as Brazil. In contrast, in countries with relatively flat income distributions, like Croatia or Armenia, pupils from different social classes are more likely to attend the same schools. This could reduce the impact of family wealth on test scores.

How Big Macs can measure the true size of China's economy

America's economy did not exceed China's in size until the 1880s, according to the Maddison Project at the University of Groningen. The two now rival each other again. Because China's workers are 4.7 times as numerous as America's, they need be only a fraction as productive to surpass America's output. No fewer than 53 countries would already have a bigger GDP than America if they were as populous as China.

In 2019 China's workers produced over 99trn yuan-worth of goods and services. America's produced $21.4trn-worth. Since 6.9 yuan bought a dollar in 2019, on average, China's GDP was worth only $14trn when converted into dollars at market rates. That was still well short of America's. But 6.9 yuan stretches further in China than a dollar goes in America. One example is the McDonald's Big Mac. It costs about 21.70 yuan in China and $5.71 in America, according to prices collected by *The Economist*. By that measure, 3.8 yuan buys as much as a dollar. But if that is the case, then 99trn yuan can buy as much as $26trn – and China's economy is already considerably bigger than America's.

Motivated by this logic, *The Economist* has compared the price of Big Macs around the world since 1986. The result is a rough gauge of the purchasing power of currencies. It suggests that many currencies are undervalued, relative to the dollar, on the foreign-exchange markets. A few, such as the Swiss franc, are overvalued. Lebanon's pound was undervalued until inflation took off in late 2019, raising local prices even as the pound remained pegged to the dollar. The Big Mac alone jumped 38% in price.

Every few years the World Bank embarks on a vastly more systematic effort to gauge purchasing power by comparing thousands of prices across the world. The results can be startling. Its survey of prices in 2011, released in 2014, showed that things in China were cheaper than previously thought and its economy was therefore much larger. Based on these estimates, the IMF calculated

The Big Mac index

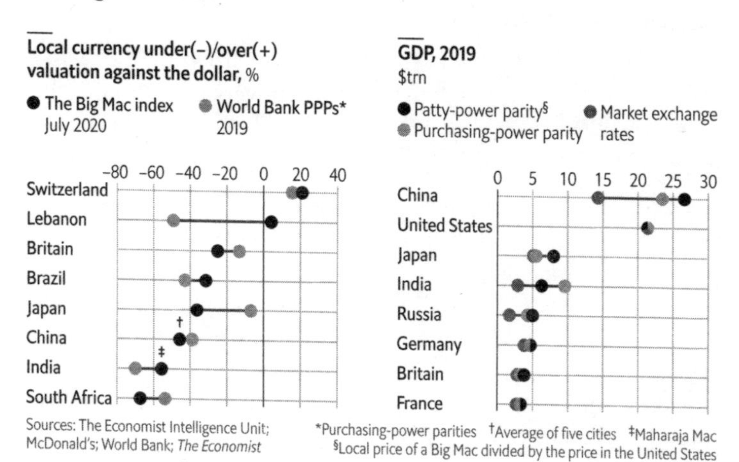

that its GDP overtook America's in 2014 and was 27% bigger in 2019. Many observers, however, greeted these estimates with scepticism. In 2010 an informal survey by a reporter at *Caixin*, a financial magazine, noted that a number of items were dearer in Hangzhou than in its sister city Boston. (It literally compared apples to apples, and found that the Golden Delicious variety was 37% pricier in the Chinese city.)

The sceptics won some vindication in May 2020 when the World Bank released its latest price-comparison exercise. It discovered that things were about 17% more expensive in China, relative to America, than previously thought. At a stroke, China's GDP fell by over $3.2trn. The estimates suggest China did not overtake America's economy until 2016.

But were these new estimates any more robust than earlier efforts? Comparing prices across the world is fraught with difficulties. An item may be a staple in one place and a delicacy in another. The World Bank must also decide how much weight to give each item. That depends on shopping habits, which differ – partly because prices differ. It is easy to go around in circles.

That is why it can help to check the World Bank's results against a cruder yardstick – like the price of a Big Mac. Our index suggests that the bank's 2020 figures, if anything, underestimate the buying power of China's currency, and therefore its economic size. McDonald's was once a symbol of America's economic might. Now the Big Mac shows how its might is being surpassed.

What bank-robbery looks like in the digital age

When Willie Sutton, a notorious 20th-century American thief, was asked why he robbed banks, he reputedly replied: "Because that's where the money is." These days there is no need to don a mask and wave a gun. Bandits can steal millions from their armchairs, wielding nothing more lethal than malicious code.

Cyberthieves grow ever more persistent, with banks and cryptocurrency exchanges among the main targets. One estimate, from 2018, put total cybercrime revenue at $1.5trn or more a year (including not only bank jobs but also theft of intellectual property, counterfeiting, data-ransoms and so on). It is likely to have risen since then, thanks to covid-19. Many financial firms struggled to keep security tight with so many staff working from home. And technically inexperienced people adopting online banking for the first time, as a result of lockdowns, provided easy targets for scammers.

Most big cyberheists are carried out either by organised-crime groups or state actors. Focus on the latter has sharpened since hackers thought to be linked to the North Korean government stole $101m (and almost got away with another $850m) from Bangladesh's central bank in 2016, after manipulating transfer instructions from SWIFT, a global payments network. After a lull in activity, the North Koreans made a comeback in 2020. American government agencies issued an alert warning that the North Koreans were cooking up new bank-robbing schemes to help fund the regime of Kim Jong Un, cash-strapped by sanctions.

One such scheme, known as an "ATM cash-out", was described by SWIFT and the financial-consulting arm of BAE Systems, a defence contractor, in a report on how cyberheists are carried out and the gains laundered. Far from being all-digital, it involves hacking cash machines to make them spit out notes, which are then grabbed by accomplices known as "money mules" who are loitering nearby. Among those who specialise in cash-outs are the BeagleBoyz, a group linked to the Reconnaissance General Bureau, a North Korean spy agency, who have attempted to steal nearly $2bn since 2015, according to the American agencies.

The downside of attacking ATMs is that each one holds limited amounts of cash. So the hackers do it in volume. In some cases, machines in more than 30 countries (including America) have been targeted in a single strike. This makes the process labour-intensive: an attack on one bank, by a group called Lazarus, involved 12,000 ATM withdrawals across 28 countries, all made within two hours, according to the SWIFT/BAE Systems report. This requires large numbers of mules to be in position at exactly the right time, as each ATM spits out its haul. The mules are also used to get the stolen cash back into the financial system. A common way to do this is to take it to a casino, convert it into chips, and then exchange it back into cash in the form of a cheque from the casino showing a legitimate transaction. This can then be deposited in a bank without ringing alarm bells.

Efforts to foil cash-out schemes tend to focus on identifying mules from CCTV footage, then trying to connect dots up the chain of command. But some are adopting cruder defensive measures. After being warned about the latest threat from North Korea, some Bangladeshi banks opted to shut down their ATMs between midnight and 6am to reduce the threat from Willie Sutton's digital-age counterparts.

Why Zoom meetings are so dissatisfying

Readers of a certain age will remember when long-distance calls were expensive, international calls ruinously so, three-way calls exciting and video calls the stuff of science fiction. How quickly people take yesterday's achievements for granted. Today, international video hangouts are free and widely available. Yet instead of treating them as a miracle, endless commentators have complained about "Zoom fatigue". Much of their criticism has been about the video: a lack of eye contact, self-consciousness (whether about skin, hair or bookshelves) and the like.

Yet the main reasons that Zoom conversations are draining are to do with audio, where the limitations of the technology run up against habits of speech. Studies show that most cultures observe a rule of "no gap, no overlap" in conversation. Turn-taking is well organised and almost instantaneous, from Mexico to Denmark to Japan. But this rule is disrupted in online meetings. Audio and video are chopped into tiny pieces, sent via different channels to the recipient, and then reassembled. When some packets arrive late, the software has a basic choice: to wait, leading to a delay, or to output what is available, leading to glitches.

Video-calling platforms tend to use audio that arrives quickly but is of middling quality. Zoom says it aims for, and often achieves, a lag of just 150 milliseconds – quicker than the blink of an eye. Yet even when that goal is reached (and it often isn't), that is a lot more time than it seems. Under "no gap, no overlap" rules, the typical silence between the end of one face-to-face conversational turn and the next is about 200 milliseconds. The wait easily exceeds that threshold if Zoom users experience a 150-millisecond lag after the first speaker, followed by another 150 milliseconds for the reply.

Adding these pauses to work calls can make speakers seem less convincing. A study by Felicia Roberts of Purdue University and colleagues found that positive answers to questions (such as "Can you give me a ride?", "Sure") were rated as less genuine if the responder took more than 700 milliseconds to reply. Above that limit, hearers perceive that the speaker is using extra time to craft a response, perhaps a hedge or a polite "no". Unfortunately, this means that colleagues who think they are giving forthright answers might come across as cagey on video calls.

A bigger problem may be interruptions, says Ms Roberts, as delays mean that speakers are not able to properly time their turns. In person, when two people overlap one speaker may quickly yield; on a video call it takes longer for this clash to be resolved. Repairing these snags regularly is tiresome.

To make matters worse, colleagues who are hard to understand, even if only for technical reasons, are rated as less trustworthy. Studies find that a foreign accent reduces the believability of factual assertions (such as "a giraffe can go without water longer than a camel") – as does printing such statements in a fuzzy or low-contrast font. In humans' primitive psychology, the easier something is to understand, the easier it is to believe. This same bias unfairly punishes workers cursed with dodgy internet connections. There is at least one upside to all this, though. As workers return to offices, they may actually look forward to real face-to-face meetings again – to say nothing of post-work gatherings with friends.

Quantifying the size of the Bank of Mum and Dad

Most young people these days need a boost to get on the property ladder. In Britain more than 60% of homebuyers under 35 get financial help from their parents, according to Legal & General, a British insurer. Such assistance averaged £24,100 ($30,000) in 2019, up from £17,500 in 2016. The so-called Bank of Mum and Dad is, in effect, the 11th-largest mortgage lender in the country. The same applies elsewhere. In America, parents are the seventh-biggest lender, according to another study.

But for many young adults, the financial support does not end there. In a paper published in June 2020, researchers at the University of Copenhagen combined customer data from Danske Bank, Denmark's biggest lender, with government records about where these customers live, work and went to school, and how much they earn. With the resulting database, the researchers tried to understand how much young Danes, aged 20 to 39, are supported financially by their friends and family.

Unsurprisingly, parents turn out to be by far the biggest backers. Young Danish adults receive a net transfer of almost $65 per month from their parents, on average (parents give an average of around $140 and receive an average of $75 from their children). But when those who neither receive money from their parents nor give them any are excluded from the sample, the average net amount rises to more than $100 per month. One young adult out of every 25 in the database was found to receive more than $1,000 per month. (The paper reported transfers in dollars, not Danish kroner.)

Young people rely on their parents the most when their income falls. Those on the middle rung of the earnings ladder receive around $50 a month net from their parents on average. When an individual falls from the median earnings bracket to the bottom 5% of earners, the amount increases by over $100. Conversely, those who move into the top 5% of earners receive $30 a month less. And parents provide more than just financial support. The researchers

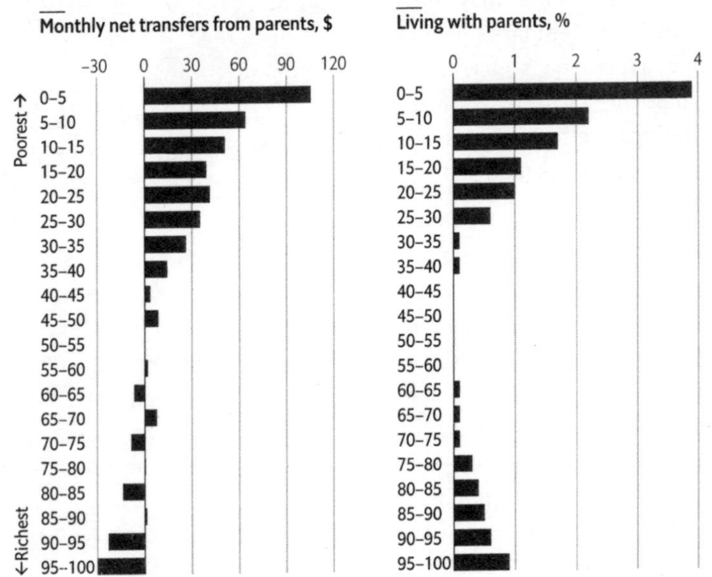

found that falling into the bottom income bracket increases the chance that a young person will live with their parents by 4% (again, compared with the median earner). Counterintuitively, young Danish people making big bucks are more likely than some of their lesser-paid peers to move in with their parents. But the authors reckon that this is evidence that rich young people are hosting their parents, rather than the other way round.

Not surprisingly, parents tend to be more generous than others in a person's social circle. Siblings, grandparents, school friends and colleagues give on average less than $20 a month combined, gross, and almost nothing net. Nor do they help out much more when things get tough. Falling from the median income bracket into the bottom one will only elicit an extra $5 a month from siblings, on average.

These results may not be representative of experiences elsewhere. Denmark's generous social-insurance policies already offer a relatively high level of income support. Parents in countries where the state provides less help may be even more generous. The findings are nevertheless clear: when young people struggle for money, the bank (and bedrooms) of Mum and Dad are the first place they turn to.

What bosses can learn from a hostage negotiator

"When people talk, listen completely." Those words of Ernest Hemingway might be a pretty good guiding principle for many managers, as might the dictum enunciated by Zeno of Citium, a Greek philosopher: "We have two ears and one mouth, so we should listen more than we say." People like being listened to. Some firms use a technique known as a "listening circle" in which participants are encouraged to talk openly and honestly about the issues they face (such as problems with colleagues). In such a circle, only one person can talk at a time and there is no interruption. A study cited in the *Harvard Business Review* found that employees who had taken part in a listening circle subsequently suffered less social anxiety and had fewer worries about work-related matters than those who had not.

Listening has been critical to the career of Richard Mullender, who was a British police officer for 30 years. Eventually he became a hostage negotiator, dealing with everything from suicide interventions to international kidnappings. By the end of his stint in uniform, he was the lead trainer for the Metropolitan Police's hostage-negotiation unit. When he left the force in 2007, he realised that his skills might be applicable in the business world. So he set up a firm called the Listening Institute. Mr Mullender defines listening as "the identification, selection and interpretation of the key words that turn information into intelligence". It is crucial to all effective communication.

Plenty of people think that good listening is about nodding your head or keeping eye contact. But that is not really listening, Mr Mullender argues. A good listener is always looking for facts, emotions and indications of the interlocutor's values. And when it comes to a negotiation, people are looking for an outcome. The aim of listening is to ascertain what the other side is trying to achieve. Another important point to bear in mind is that, when you talk, you are not listening. "Every time you share an opinion, you give out information about yourself," Mr Mullender says. In contrast, a good listener, by keeping quiet, gains an edge over his or her counterpart. Hostage negotiators usually work in teams, but the lead negotiator is the only one who talks. "What we teach is that the second person in the team doesn't really talk at all, because if they are busy thinking about the next question to ask, they aren't really listening," Mr Mullender explains.

The mistake many people make is to ask too many questions, rather than letting the other person talk. The listener's focus should be on analysis. If you are trying to persuade someone to do something, you need to know what their beliefs are. If someone is upset, you need to assess their emotional state. Of course, a listener needs to speak occasionally. One approach is to assess what the other person is telling you and then check it with them ("It seems to me that what you want is X"). That gives the other party a sense that they are being understood. The fundamental aim is to build up a relationship so the other person likes you and trusts you, says Mr Mullender.

As a result of the pandemic, far more business conversations now take place on the phone or online. Precious few in-person meetings occur. Some might think this makes listening more difficult; it is harder to pick up the subtle cues that people reveal in their facial expressions and body language. But Mr Mullender says that too much is made of body language. It is much easier to understand someone if you can hear them but not see them, than if you can see but not hear them. He prefers to negotiate by telephone. Remote working has increased the need for managers to listen to workers, because the opportunities for casual conversation have dwindled. Mr Mullender thinks that many people have become frustrated in their isolation, which can lead to stress and anger. He thinks there may be a business opportunity in helping managers listen more efficiently, so they can enhance employee well-being. After months of isolation, many workers would probably love the chance to be heard.

Globally curious: particular propensities from around the world

Why Corsican number plates are so popular in France

Just over a decade ago, France abandoned a rule that had obliged motorists to change their car's number plate each time they moved house to a new administrative *département*. The point was to ensure, in true bureaucratic style, that the vehicle's plate matched the owner's place of residence. Since 2009, however, car owners have been free to choose which *département* code they display, turning number plates into a test of sentimental attachment – with unexpected results.

The surprising favourite is Corsica, an island that is home to just 340,000 people. The 2A that represents one of the island's two *départements*, along with its symbol of a bandanna-wrapped head, was the most sought-after plate, relative to the local population, over an eight-year period. Mountain regions were also popular. Paris did not get into the top ten.

This *amour* for Corsica, which became part of France in 1796 after a history of contested independence and conquest, may simply reflect the strong regional identity of the island's diaspora. To display a 2A or 2B *plaque* is a badge of pride and belonging. Perhaps, suggest some, it is also a declaration of love for the "isle of beauty", as it is known to French tourists, 2m of whom flock there in a normal year.

But another, more likely possibility is that a Corsican number plate, consciously or not, is regarded as a form of implicit protection. The island has a history of violent nationalism, after all. The Corsican National Liberation Front waged a decades-long armed campaign for independence. No mainlander wants to attract too much attention there. Two decades ago Claude Erignac, the *préfet*, or central-government representative, was assassinated. Mob and clan rule on the island have a broad hold on the French imagination. Honoré de Balzac's novel, *La Vendetta*, was about Corsican vengeance and family honour.

Even non-Corsicans in Paris are known to drive with Corsican plates. "They are convinced that their car won't be damaged, and that they won't be bothered on the roads," suggests Benoit Ginet, founder of Eplaque.fr, an e-commerce number-plate business, which registered a disproportionate demand for plates from the island in 2020. There may also be a form of superstition at work. "Motorists feel stronger with Corsican plates," he suggests. Whether to other drivers, or to the fates, Corsican plates send a signal: don't mess with me.

Why tea sales have boomed in Johannesburg's restaurants

Upon arrival at an expensive restaurant near Sandton, Johannesburg's financial district, nothing feels unusual, at least in the era of covid-19. The maître-d' zaps a temperature gun at patrons' masked faces and spritzes their hands with sanitiser. Only when ushered into the dining room does something seem odd: every table has a pot of tea. South Africans like their *rooibos*, but the liquid in many tea-cups is a darker red – one might call it claret, even. On other tables there are cans of tonic water next to teapots. The penny drops when two young women are given shot glasses of Jägermeister inside their tea-cups.

South Africa banned alcohol sales for the second time in 2020 on July 12th, to prevent drunks from taking up precious space in hospitals. But restaurants, which are struggling to stay afloat, remained open and customers were thirsty. "Are you serving special teas?" your correspondent hesitantly asked a waiter. "Yes," he replied, the ripple of a smile just visible from the edges of his mask. An unscientific sample of local establishments highlighted the patchiness of prohibition. A steak house in an affluent suburb also used teapots for red wine, and put cans of fizzy apple juice next to glasses of beer. Other joints were more brazen. At a café popular with families, parents kept one eye on their children ricocheting inside a bouncy castle and the other on their ice buckets.

The willingness of South Africans to flout the booze rules was a reminder that bans rarely work. The police were unwilling or unable to enforce them – at least in rich areas. But it also reflected how few South Africans are willing to obey laws set by a ruling party that itself shows little respect for propriety. In mid-2020 local journalists revealed how members of the African National Congress and their families had won juicy contracts for supplying personal protective equipment, despite having little experience in medical kit. Cronyism has been rife for years, but the sight of "tenderpreneurs" coining it was especially galling when thousands were dying. On August 3rd President Cyril Ramaphosa compared such people to hyenas circling wounded prey.

Yet he has done too little to stop the scavengers. More than two years after Mr Ramaphosa took office, South Africans were still waiting for prosecutions to be brought against those accused of wrongdoing during the reign of his predecessor, Jacob Zuma. For his part, the former president continued to stall his own trial over an arms deal cut more than 20 years ago. It is enough to make one want to order a strong cup of tea.

Why self-help books are so popular in China

Bookshops in China are replete with works offering advice on self-betterment. Topics range from coping with shyness ("How to Make Friends with Strangers in One Minute") to succeeding in business ("Financial Management in Seven Minutes"). The title of one bestseller urges: "Don't Opt for Comfort at the Stage of Life that is Meant to be Difficult". These books' popularity and contents reflect the stresses of a society in rapid flux – one in which paths to wealth are opening up in ways barely imaginable a generation ago, and where competition is fierce. Reliable statistics on China's book market are hard to find. But according to a study by Eric Hendriks-Kim, a sociologist at the University of Bonn, self-help books may account for almost one-third of China's printed-book market. In America they make up only 6% of adult non-fiction print sales, reckons NPD Group, a research firm.

Although China's leaders keep stressing the need for China to be "self-reliant", seekers of advice on how to succeed often turn to American books for guidance. In China in 2019 the top ten self-help sellers included translations of several American works, such as *How to Win Friends and Influence People*, *Peak: Secrets from the New Science of Expertise* and *The Seven Habits of Highly Effective People*. Chinese readers appear more eager for such imports than people in many other countries that are culturally closer to America. That may be because both China and America are "hyper-competitive and materialistic regimes", argues Mr Hendriks-Kim, who has described this phenomenon in his own book, *Life Advice from Below: the Public Role of Self-Help Coaches in Germany and China*. In the early 2000s a Chinese translation of *Who Moved My Cheese?*, a motivational book by an American, Dr Spencer Johnson, became so popular that a play based on it toured theatres and the Chinese word for cheese acquired a new meaning: one's own self-interest. Books with "cheese" in their titles proliferated in China.

China has a long tradition of reading for practical purposes. In 2018 fiction accounted for 7% of sales, compared with more than

30% in Germany. "One of the most striking features of China's market for books is its absolute and passionate relevance to life," noted a report in 2006 by Arts Council England. The exam-focused education system leaves little time to develop interpersonal skills, so people, desperate for advice on how to sell themselves, turn to self-help books instead.

That may suit the Communist Party, eager as it is to promote "positive energy". But the party would prefer native-born role models. State media have touted a book by President Xi Jinping, *Seven Years as an Educated Youth*, as the kind of tome people should study. It describes Mr Xi's hard life in the countryside during the Cultural Revolution of the 1960s and 1970s. "Is there really any self-help book better than Xi's?" asked one headline. Mr Xi is also fond of the classics, some of which are being repurposed for self-improvement purposes. Yu Dan, perhaps China's best-known pop philosopher, has sold 11m legal copies (millions more may have been peddled in photocopied form) of *Confucius from the Heart*. Some Chinese have mocked it for making the sage sound "much like the masters of American self-help", says Mr Hendriks-Kim.

Perhaps the self-help industry has come full circle. After all, China's 6th-century-BC masterpiece for would-be generals, *The Art of War* by Sun Tzu, was arguably the self-help prototype. Its title has been echoed, consciously or otherwise, in the names of countless other books of the genre. One such is Donald Trump's *The Art of the Deal*. Its fifth and most recent translation in China was published in 2016 – by the Communist Youth League.

Why the Baltic states are reconfiguring their electric grids

It doesn't sound dramatic. Technicians in the Baltic countries of Estonia, Latvia and Lithuania are preparing to change the frequency of their electric grids. This will involve desynchronising from a regional power system called IPS/UPS to allow synchronisation with another one, the Continental Synchronous Area. But look closer, and the switch is part of a contest that pits democratic Europe against autocratic Russia and its tinpot ally Belarus.

One legacy of the Baltic states' past as involuntary members of the Soviet Union is that the mains frequency of their IPS/UPS power system is controlled from Moscow. This means Russia's regime could switch off the Baltics' power for the better part of a dark and possibly cold week, or perhaps longer, while Baltic operators scramble to restore power with local means. The first three days alone of such a blackout would cost the Baltics €2.3bn ($2.7bn) in lost output, reckons Taavi Veskimagi, boss of Elering, Estonia's grid operator. Deaths and social unrest could add to the toll. Russia has not explicitly threatened a Baltic blackout. The Kremlin has, however, occasionally cut off hydrocarbon exports, just to remind eastern Europeans who's in charge. Russia could add grid power to its "strategic coercion" repertoire, especially if political upheaval led its leaders to seek support by manufacturing a crisis abroad, says Tomas Jermalavicius, formerly a planner at Lithuania's defence ministry.

A big outage in Latvia on June 9th 2020 concentrated minds. Nothing indicates that the central dispatch office in Moscow was behind it. Even so, authorities there, says Mr Jermalavicius, "just sat on their hands and watched" instead of stepping in to prevent more cascading blackouts. (An emergency inflow from Poland saved the day.) At the least, then, Russia appears disinclined to help in a pinch. So the Baltic states must synchronise as quickly as possible with Europe, a "trusted area of high standards and legal norms", says Mr Jermalavicius, now head of studies at the International Centre for Defence and Security, a think-tank in Tallinn, Estonia's capital. Hoping to reduce this threat to its north-eastern flank, the EU is expected to cover three-quarters of the project's whopping cost of €1.6bn. America is also chipping in. Its departments of energy, state and defence, as well as the CIA, are providing money, kit and expertise. Pro-Kremlin propaganda urging people in the three states to oppose the grid reconfiguration has failed to gain traction. Even so, the project will not be complete until 2025.

Baltic strategists have long taken comfort in one aspect of the regional power system's design. Were the Kremlin to trigger a blackout in the Baltic states, power would also go out in areas of Belarus and western Russia, due to the synchronous cross-border connections. But this restraint on the Kremlin's options is slipping away. Russia is rapidly reconfiguring and upgrading its own grid in a way that will insulate itself and Belarus, even if the Baltic states go dark. A recent development concerns Kaliningrad, a heavily militarised Russian exclave detached from the mainland and sandwiched between Lithuania and Poland. Edvinas Kerza, a former vice-minister of defence, says that Lithuanian intelligence has determined that in late 2019 Kaliningrad's technicians achieved the ability to operate their grid even if power is down in the Baltic states. Mainland Russia will probably be insulated from any Baltic blackout by the end of 2021. That is well before the region will be ready to pair with the Continental Synchronous Area.

As a result, a quiet but high-stakes race is under way. The Baltic states are upgrading their infrastructure to shorten the time that a locally managed grid reboot would take, says Zygimantas Vaiciunas, Lithuania's former energy minister. Thanks in part to EU funds that have already been allocated, recent progress in Lithuania has probably prevented any shutdown from becoming a national disaster, he says. If a kill switch were to be flipped in Moscow, Mr Vaiciunas reckons that Lithuania could restore its grid in four days or less. That could still make for a pretty chilly episode.

Why ant-egg soup is a nostalgic treat in Laos

You wouldn't serve this to the king," says Dalaphone Pholsena, a restaurateur in Vientiane, the country's capital. Before her are two small bowls of ant-egg soup, a favourite dish of the late summer in Laos. In it are chunks of white fish, meaty mushrooms and dozens of splayed and lifeless ants. Bobbing on the surface is the *pièce de résistance*: clusters of ivory-white eggs that look like tiny white beans. They burst in the mouth like fish roe, but with a more acidic tang.

Ms Dalaphone considers herself a defender of the dishes traditionally eaten by Laos's subsistence farmers – of which there are still many. Ant-egg soup is a classic: both an important source of protein and an emblem of rural life. The eggs are laid by red weaver ants, which nest in mango trees and coconut palms in April and May. A brave forager – the ants' bites are like the prick of a needle – uses a stick to tear open the nest, catching egg-covered leaves (and lots of livid ants) in a bucket. Wearing as few clothes as possible, the better to brush marauding ants from the skin, and hopping from foot to foot to evade the enraged insects, the harvester then shakes the eggs from the leaves. Rural folk tend to mix this hard-won prize into omelettes, salads or soups, adding a distinctive, sour pop.

Nowadays, however, many pack up the eggs and ship them to markets in nearby cities instead. Laos is urbanising fast. In 2000 about a fifth of its population lived in cities; today over a third does. New urbanites often express nostalgia for the countryside. On weekends many middle-class Laotians drive to the family village to help tend the rice paddies. Failing that, a steaming bowl of ant-egg soup can be almost as transporting. "It's something that reminds them of life before," says Gie, a 30-something professional. "You think of 50 or 20 years ago, when you were in the village, with your mum."

A kilo of eggs can fetch as much as 150,000 kip ($16) in Vientiane. That is a handsome sum for a poor rural forager. But even as the price of eggs climbs, people like Ms Dalaphone worry about the dish's future. Urban youth grow up eating pizza and wontons and are often squeamish about gulping down bugs. Asked about ant-egg soup, Gie's 12-year-old son replies, "It looks awful, not a tasty meal." He prefers fried rice and noodles, easily ordered by phone or whipped up in an instant from packets imported from China or Thailand.

Chefs say traditional Lao cuisine, including ant-egg soup, needs a charm offensive to survive. A reprieve may come from the covid-19 pandemic, which has squeezed incomes and pushed some people to revert to cheaper folk dishes, or even to foraging, to save money. But in the long run, in all likelihood, fewer and fewer Laotians will be willing to brave a sting for their supper.

Why is control of Western Sahara so controversial?

President Donald Trump's decision in December 2020 to recognise Morocco's annexation of Western Sahara – in return for Morocco's decision to establish diplomatic relations with Israel – turned up the heat on a conflict that had been simmering for four decades. On Twitter, Mr Trump announced that "Morocco recognized the United States in 1777. It is thus fitting we recognize their sovereignty over the Western Sahara." But the annexation is not accepted by most countries, nor by the Polisario Front, which fought Morocco for control of the territory from 1975 to 1991. Polisario, which still controls a third of Western Sahara, is recognised by the United Nations as the legitimate representative of the indigenous Sahrawi people. So who should control Western Sahara?

Once known as Spanish Sahara, the area was the last vestige of Spain's colonial empire. When Spanish troops withdrew in 1975, it became a battlefield with Morocco, Mauritania and Polisario laying claim to the area, rich in phosphate deposits and fish stocks. Although Mauritania long ago gave up on its claim, withdrawing from the territory in 1979, these economic interests remain an important part of what drives the conflict. Morocco asserts a historical right to control the region dating back to before the Spanish arrived. But the Sahrawi people insist that they have largely governed themselves for hundreds of years, roaming the desert as nomads with ties to no kingdom or state beyond their own tribal allegiances. Neither 16 years of fighting, nor the past 30 years of ceasefire and talks between Morocco and Polisario, has resolved anything.

The UN tried to help the former colony make a smooth transition from Spanish rule. Several resolutions, passed in the 1960s and 1970s, emphasised that all powers should be transferred to the people being decolonised, "in accordance with their freely expressed will and desire". This meant consulting the Sahrawis in a referendum. Did they want to join one of their neighbours (Morocco or Mauritania), or seek independence? But the referendum hit an immediate snag when Morocco asked the International Court of Justice to consider its claim over the territory. The court found that Western Sahara was not a *terra nullius* (nobody's land) at the time of Spain's colonisation and belonged to someone – and that strong historical ties connected Morocco to the tribes living in Western Sahara. But did this mean it could annex the territory hundreds of years later? Morocco claimed as much, and the court's decision continues to form the basis of its claim today.

To show that Western Sahara formed part of its national territory, Morocco had to produce evidence of authority at the time of Spain's colonisation and immediately before it. That its then sultan had appointed *caids* (tax collectors) and sheikhs in the territory in the 19th century was not enough to convince the judges. According to Spain, a notable characteristic of the tribes in the territory where the sultan had nominal religious and political control was their refusal to be taxed, which made Moroccan authority seem shaky at best. The judges agreed that the sultan displayed authority over some of the tribes in Western Sahara, but they rejected any other "tie of territorial sovereignty".

After a round of talks in Madrid, also in 1975, Spain agreed to transfer its responsibilities as the administering power to Morocco and Mauritania. Shortly afterwards, Morocco began to "peacefully occupy" the territory. The Sahrawis responded by forming Polisario, which was backed by neighbouring Algeria – Morocco's long-standing rival for control of the routes to west Africa. Continued Algerian support for Polisario since 1975, through the provision of financial aid, arms and training, has ensured the liberation movement's survival.

But the historical argument has proven largely irrelevant. A war between Morocco and Polisario ended in a ceasefire in 1991, with Morocco in control of about two-thirds of the territory and all its coastline, and Polisario running about a third, on the border with Mauritania. No referendum has taken place, and as time goes on the rationale for holding one weakens. The Moroccan government has steadily increased the presence of Moroccan settlers in Western Sahara, ensuring its authority by sheer numbers. Determining who should be entitled to vote in the referendum has therefore resulted in a long deadlock. The original list for voter identification, agreed to by both Morocco and Polisario, says that only citizens featured in the 1974 census of the Western Sahara population and their descendants should be eligible. Morocco wants to let its citizens vote in any referendum. And with Moroccan settlers now accounting for some 80% of Western Sahara's population, this would lead to a predictable outcome. But even this is unlikely. Morocco has already spent billions developing the region and has no interest in holding a referendum. If Morocco has its way, Western Sahara will be autonomous, not independent. America's recognition of the *fait accompli* has made that all but certain.

Where are the world's most expensive cities?

Osaka's reputation as "the kitchen of Japan" stems not from its culinary prowess but from the many warehouses that have stored the nation's rice and other goods for centuries. Osaka can now also claim the title of the world's most expensive city, according to the 2020 findings of the Worldwide Cost of Living survey, compiled by The Economist Intelligence Unit. Osaka shares the top spot with Hong Kong and Singapore, both previous holders of this dubious honour.

The survey, which was conducted in late 2019, before the spread of covid-19, compared the prices of more than 150 items in 133 cities around the world. The results are primarily used by firms to negotiate appropriate compensation when relocating staff. The three Asian cities that lead the ranking were found to be more expensive than New York, the benchmark, albeit by just 2%. Paris, which shared the top spot with Hong Kong and Singapore last year, fell four places to fifth. Indeed, of the 37 European cities surveyed, 31 experienced a fall in overall rank because modest domestic demand

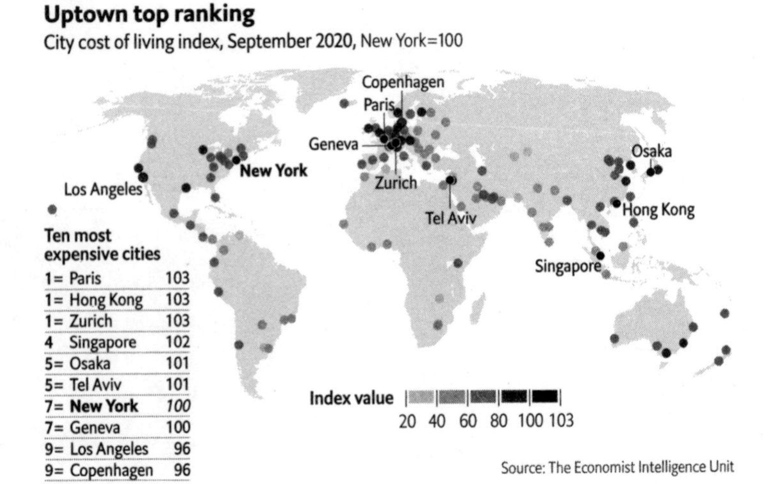

and weak global energy prices kept inflationary pressures subdued across the region.

In contrast, America's strong economy and the appreciation of the dollar pushed American cities up the rankings. Fifteen of the 16 in the survey rose. The highest climbers were Boston (from 51st to 33rd), where fast-paced population growth continued to push up demand; Atlanta (76th to 63rd), where a strong jobs market and rising house prices had a similar effect; and San Francisco (25th to 15th), where a high concentration of wealthy individuals meant that prices continued to rise quickly for more luxurious goods and services in categories such as personal care, clothing and domestic help (for which it is the most expensive city in the world).

The lower end of the ranking, by contrast, is populated by cities facing political and economic instability. The Syrian capital, Damascus, afflicted by civil war (and hence soaring inflation and goods shortages), returned to the bottom of the index.

What did Lava Jato, Brazil's anti-corruption investigation, achieve?

The rollers and spray jets have largely stopped, but Brazil's politicians and officials are far from squeaky clean. Lava Jato ("Car Wash") started in 2014 as an unremarkable money-laundering probe before growing into a much wider inquiry into malfeasance. But in early 2021 the task force of prosecutors who had led the probe in Curitiba, in southern Brazil, was wound up. The public was too distracted by the covid-19 pandemic and the terrible state of the economy to take much notice. The inquiry led to the conviction of 174 people, including a few of Brazil's most senior politicians, and recovered at least 26bn reais ($5bn) for the public coffers. But on March 8th 2021 Luiz Inácio Lula da Silva, a popular former president and the investigation's prize catch, had two convictions for corruption annulled by Edson Fachin, a Supreme Court justice. So did Lava Jato achieve lasting change?

Lava Jato's name belies its seriousness. The name arose because the original probe focused on a petrol station in Brasília, the capital, which was used by money-launderers. Ironically, it did not have a car wash. One black-market money-dealer was found to have links with Paulo Roberto Costa, a former director of Petrobras, the state-controlled oil company. It emerged that Petrobras had been awarding over-generous state contracts to construction companies, which then ensured that Brazil's politicians and their parties got their cut. The inquiry expanded as investigators uncovered a mass of related offences, and hundreds of arrests followed. Prosecutors, overseen by a young judge, Sérgio Moro, made heavy use of pre-trial detention to extract plea bargains, then a new weapon in the judiciary's arsenal.

The revelations of Lava Jato brought Brazilians onto the streets in protest and made them believe the country's rampant corruption could be overcome. Politicians of all stripes were caught up, among them two of the three immediate predecessors of the current president, Jair Bolsonaro. Lula served 18 months of a 12-year term for corruption and money-laundering before being released. Michel

Temer was accused of embezzling 1.8bn reais and remains under investigation. Though Dilma Rousseff, whose presidential term separated those of Lula and Mr Temer, was not directly implicated in the scandal, Lava Jato played a role in her impeachment. She had chaired the board of Petrobras when many of the bribes happened and became deeply unpopular. The charges Ms Rousseff faced related to her manipulation of the federal budget to disguise Brazil's economic problems, but it is widely thought that Eduardo Cunha, the speaker of the lower house at the time, initiated impeachment proceedings because Ms Rousseff refused to protect him from the investigation. Mr Cunha was convicted of corruption, tax evasion and money-laundering.

Lula's legal troubles were enough to keep him out of the 2018 election, won by Mr Bolsonaro. The new president made Mr Moro justice minister, but Mr Moro's reputation for impartiality was soon further tarnished. In June 2019 a trove of hacked messages revealed that he had collaborated with prosecutors, offering advice and investigative leads. Suspicions arose that Mr Moro was motivated less by a concern for justice than political animus. Mr Fachin quashed Lula's convictions on the grounds that Mr Moro's court in Curitiba lacked jurisdiction. Although the case is likely to be sent to another court, and Lula faces other charges, this freed him to run for president again in 2022.

Mr Moro's fall from grace fuelled a backlash among the political and judicial classes. The Supreme Court and other tribunals, which at first supported the investigation, started ruling against it and in favour of politicians as the investigation's net expanded. Mr Bolsonaro lost whatever interest in tackling graft he might have once had when his own family was caught up in a separate corruption investigation. He declared that Lava Jato would wrap up, since there was "no more corruption in the government". Mr Moro resigned as justice minister in April 2020, claiming that Mr Bolsonaro sacked the chief of the federal police to protect his sons, Flávio and Carlos, both of whom are under investigation. In February 2021, the task force of prosecutors in Curitiba was disbanded.

Any success that the investigations, arrests and convictions had in cleaning up Brazilian politics will probably be ephemeral. Lava Jato removed some bad actors from the stage, but failed to fix the culture of corruption. It pioneered new techniques, and a new boldness in tackling corruption and the traditional impunity enjoyed by the powerful in Latin America. But its demise was a blow to anti-corruption drives elsewhere in the region. Many of those investigations are in their early stages and, like Lava Jato, subject to concerns about political control. The lasting lesson from Lava Jato is the vital importance of insulating the judiciary from politicisation.

Why western European armies have shrunk dramatically

When the cold war ended, the West celebrated the peace dividend. Armies shrank, tanks were mothballed and money for defence dried up. American military spending fell from 5.3% of GDP in 1990 to 2.9% a decade later. But in Europe demobilisation went even further. Spending in western Europe slumped from an average of 2.4% to 1.6% of GDP, according to the Stockholm International Peace Research Institute, a think-tank, and continued to slide even after al-Qaeda's attacks on America on September 11th 2001. A report released in February 2020 by the International Institute for Strategic Studies (IISS), another think-tank, shows the dramatic effect this had on Europe's military might.

In 1990 West Germany alone was able to field 215 combat battalions (a battalion typically has a few hundred soldiers, and slots into a larger brigade). By 2015, even with Germany reunited, that had fallen to 34 battalions, a remarkable 84% cut. Similarly, the number of Italian battalions fell by 67% and British ones by almost half. "To a significant degree Europeans remain dependent on US military capabilities for their defence," the IISS concluded. But American troops, who defended Europe's frontier with the Warsaw Pact in huge numbers during the cold war, went home in droves afterwards. American forces in Europe shrank from 99 battalions to a paltry 14 – from half a million troops to 76,000 (which is still more than all but seven European NATO allies).

Critics respond that such bayonet-counting is no longer relevant to modern warfare; what matters is technology, not troops. Precision weapons (as opposed to "dumb" or unguided ones) made up 6% of all munitions used in the 1991 Gulf war. That rose to 26% in the Kosovo war of 1999, 68% in America's invasion of Iraq in 2003 and 100% during the Libya war in 2011, according to figures collated by IISS. Far fewer squadrons could do the same job. During the second world war it took 1,000 or so bomber sorties to destroy one target. By the time of the Vietnam war, it took only 20 sorties

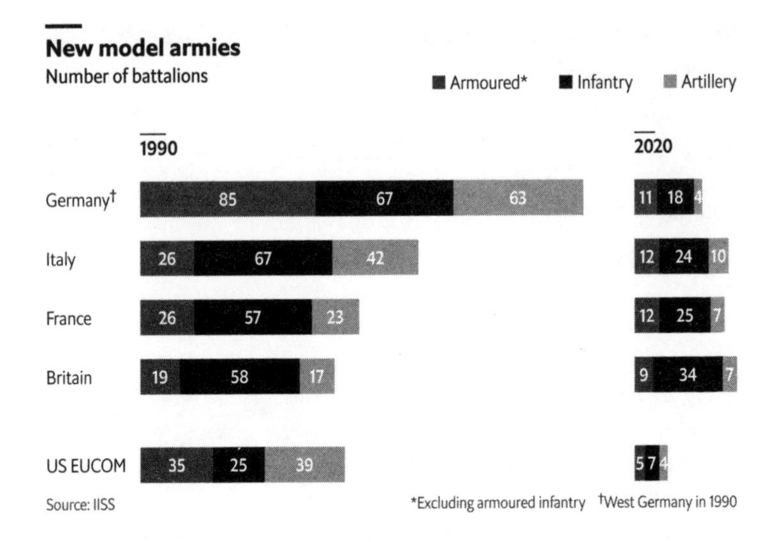

using early laser-guided munitions. By 1991 a single warplane could hit two targets with a single pair of bombs. And in Kosovo a lone B-2 bomber could take out 16 targets in a single sortie.

But numbers still matter in some circumstances, says James Hackett of IISS. Urban warfare might require large numbers of boots on the ground, for instance. And with tiny forces, small losses can have a big impact. When Norway crashed and sank a warship in 2018, its frigate fleet shrank by a fifth. Some American officials note, only half-jokingly, that Britain's army could be lost in an afternoon of combat against a serious adversary. Perhaps most important, Mr Hackett points out that Russia has modernised its own forces – replacing 1970s-vintage *Tochka-U* missile brigades with cutting-edge *Iskander* ones – without shrinking to the same extent as western European armies.

Europeans are waking up to some of this. In 2019 defence spending among EU members almost returned to pre-financial-crisis levels, though that represented just 1.4% of GDP on average, far below NATO's target of 2%. European NATO members, along with Canada, have added $130bn to their aggregate defence budgets since 2016. Germany, which accounted for a fifth of that rise, activated an additional sixth tank battalion in 2020. That was a start, but it will be a long way back to decent levels of readiness. During a Bundeswehr exercise in 2014, hard-up commanders were forced to strap broomsticks to their armoured vehicles to stand in for machine guns.

Where does Britain's royal family get its money from?

"All I wanted was enough money to get security and keep my family safe," Prince Harry told Oprah Winfrey, a talk-show host, during an interview in March 2021 alongside his wife Meghan. Other revelations from the interview, including claims of racism within Britain's royal household, explained why the Duke and Duchess of Sussex left the firm. The Sussexes have been funding their life in California with splashy media deals, producing podcasts for Spotify, an audio-streaming service, and documentaries for Netflix, a video platform. Hawking themselves to the highest bidder is a far cry from the official engagements they left behind, and it remains to be seen whether it will prove as lucrative. In 2020 the *Times* reported the queen's private wealth to be £350m ($486m), but the royal family's total assets are worth many times that. Where does Britain's royal family get its money from?

As head of state, the monarch technically owns the crown estate, a collection of land and assets that includes Ascot racecourse, a big chunk of central London and half of the foreshore (the coastal land between the high- and low-tide marks) in England, Wales and Northern Ireland. But the crown estate is run independently, and its profits – £345m in 2020 – go to the government. Some of this money is then given back to the royal household, under what used to be called the "civil list". This was worth £13.7m in 2011, most of which was used to pay staff. In addition, the monarchy also received "grants in aid" to cover travel, communications and the upkeep of palaces. In 2011 that amounted to £18.4m (£6m for travel, £11.9m for property and £500,000 for communications). In 2012, however, this funding was replaced by a single payment, the "sovereign grant". This has risen rapidly, notably in 2016 to cover a pricey, decade-long refurbishment of Buckingham Palace, which is crumbling. (In 2007 falling masonry narrowly missed Princess Anne's car.) The sovereign grant is now set at 25% of the crown estate's annual profits (it will revert to 15% in 2027), and in 2020 it paid out £82.4m.

On top of this, the family pockets the income from the duchies of Lancaster and Cornwall, which the queen and Prince Charles inherited as the monarch and monarch's heir, respectively. These private estates are made up of land, property and financial assets. In 2017 it emerged that the Duchy of Lancaster invested millions in a Cayman Islands fund. (Duchy Originals, a company set up by Prince Charles to sell expensive biscuits and other "natural food", is separate, and its profits go to charity.) In 2020 the duchies' assets were worth a combined £1.6bn, and made annual profits of around £47m between them. This money pays for the royals' private expenses, although Prince Charles pleads in mitigation that he gives some of his share to charity and pays income tax. Finally, although many of the palaces in which the royals live are held on behalf of the country, and so cannot be sold, they have other private property too. The queen, for example, inherited two country estates, Balmoral in Scotland and Sandringham in Norfolk, from her father, George VI.

The Sussexes' life as working royals would have been paid for in part by the sovereign grant. Prince Charles also spent £5.6m in the year to March 2020, funding them and the Duke and Duchess of Cambridge, and Princess Diana left the bulk of her estate, worth around £13m after tax, to her sons. Prince Harry told Ms Winfrey that without this money from his mother, the couple would not have been able to move to America.

This mix of private and government-funded income explains the royal family's wealth, but does little to answer thornier questions over the monarchy's value. Royalists will point out that, before the covid-19 pandemic at least, crown-loving tourists flocked to Britain, although the exact benefit is impossible to quantify. The queen pays income tax, even though she is not obliged to, and even council tax on Buckingham Palace. By leaving the royal household and singing for their supper, Harry and Meghan will hope to avoid such scrutiny. But the price will be an income that is certainly less secure, and perhaps less generous too.

Why Spain's high-speed trains are such poor value

Atocha station in Madrid is a daily marvel. From its platforms a phalanx of sleek white high-speed trains, known as AVEs, streak across Spain. Punctual and comfortable, they have slashed journey times in a large and mountainous country long notorious for poor transport. For Spaniards, they are an icon of modernity. But many are not worth the money spent on them, says a report published in 2020 by the independent fiscal authority (known as AIREF from its Spanish initials).

Over the past three decades Spain has poured money into transport infrastructure, including motorways and airports as well as AVEs. It now has 3,086km (1,929 miles) of high-speed rail lines (capable of supporting speeds of more than 250kph), second only to China. Passenger numbers have almost doubled in the past decade, as the AVEs have grabbed business from domestic flights. Even so, per kilometre of track, there are less than a third as many passengers as in France.

The network has cost €61bn ($72bn) so far. AIREF's report included the first thorough cost-benefit study of the trains. It found that the benefits, including to the environment, were less than the total cost, though that could eventually change for the lines from Madrid to Seville and to Barcelona. But there are plans for a further 5,654km of high-speed lines, at a cost of at least €73bn. Far better, AIREF suggested, to invest in neglected commuter networks, which carry 89% of rail users. It urged the setting up of an independent agency to set transport priorities and evaluate projects.

The Socialist-led government broadly welcomed the report. It may find it hard to implement. One problem is that some of the pending lines are half-built, such as those to Galicia and in the Basque Country. It may make sense to finish them. "The important thing is to stop new lines starting," said Ángel de la Fuente of Fedea, a think-tank. Since they would serve less-populated areas, they would have even fewer passengers.

Places that lack high-speed trains understandably feel left out. Take Extremadura, the poorest region of mainland Spain: ancient diesels take six hours to cover the 400km between Madrid and Badajoz. The covid-19 crisis means money is tight. But with Spain's Congress more fragmented than ever, the price of a vital vote has often been the promise of yet another high-speed track.

The art of the possible: politics, local and global

Why is Washington, DC not a state?

On a list of American states and territories ordered by area, Washington, DC is at the very bottom: nearly 10,000 Washingtons could fit into Alaska. Reorder that list by federal taxes paid per person, however, and Washington finishes first. But unlike residents of the 50 states, Washingtonians are unrepresented in Congress. They have no senators, and just one non-voting House delegate – Eleanor Holmes Norton, who has held the job for 30 years. Washingtonians could not vote in presidential elections until 1964. Not until 1974 were they permitted to elect their own municipal government, and the city still cannot pass laws or even an annual budget without Congressional review. Its residents are tiring of this arrangement. In a referendum in 2016, 85% of them said they wanted statehood. Why have Washingtonians been stateless for so long?

Washington replaced Philadelphia as America's capital in 1800, as a result of a bargain struck a decade earlier between Alexander Hamilton, then treasury secretary, and James Madison, a congressman from Virginia – reportedly over dinner at Thomas Jefferson's home in New York. Hamilton wanted the federal government to assume debts acquired by states during the Revolutionary War. That would have created a system of public credit, and given Hamilton power to shape the young country's economy. Madison, who like many southerners was wary of federal power, opposed the measure. But in exchange for Hamilton's support for moving the capital out of the North, and reducing Virginia's tax obligation, Madison rallied support for the treasury secretary's measure, allowing it to become law. The capital was moved from Pennsylvania to a plot of land at the head of the Potomac River, adjacent to the port town of Georgetown (now one of DC's most charming and expensive neighbourhoods).

The constitution grants Congress jurisdiction over the district designated as the seat of the federal government. In some ways this allowed DC to be ahead of its time. It outlawed slavery eight-and-a-half months before Abraham Lincoln's emancipation proclamation, and gave African-American men voting rights in 1867, three years before the 15th Amendment extended that right nationally. Yet this freedom was short-lived. In the early 1870s Congress stripped DC's voting rights to curtail the growing political power of African-American Washingtonians. John Tyler Morgan, a senator from Alabama and former confederate general, later said that Congress had "to burn down the barn to get rid of the rats... the rats being the negro population and the barn being the government of the District of Columbia".

Contemporary opponents of statehood for DC are not so blatantly racist. The main reason why Republicans oppose it today is that it would add two seats in the Senate, as well as one in the House, all of which would probably be held by Democrats (since 1964, when Washingtonians first voted in a presidential election, no Republican has won the city's three electoral votes). It would also strengthen calls for other American territories, such as Puerto Rico, to become states, which Republicans worry would further increase the influence of Democrats. Statehood would also deprive Congress of its authority over DC, which lawmakers have exerted to block policies proposed by the city council.

Attempts over the years to induct DC into the union as the 51st state have repeatedly failed, but recent events are giving the cause new momentum. In 2020 Donald Trump ordered federal forces to break up Black Lives Matter protests against the wishes of DC's mayor. It "was the physical manifestation of the District's disenfranchisement", said Ms Norton. The city's lack of authority over its national guard also meant it was slow to respond when insurrectionists assaulted the Capitol on January 6th. The Washington, DC Admission Act – which would carve out an enclave, putting federal buildings within a stateless national capital, but create the new "State of Washington, Douglass Commonwealth", where most of its residents live – passed the House in June 2020; its sister Senate bill has more co-sponsors than any previous statehood measure. President Joe Biden has also signalled his support, but with the Senate split 50-50, it is unlikely that Democrats will achieve a filibuster-proof majority for statehood. DC's advocates may find it will take a compromise, like that in 1790, to find a way forward.

Why has civil war returned to Ethiopia?

By some reckonings Ethiopia is the oldest independent country in Africa, but it is far from united. Tensions among its 80-plus ethnic groups grew during 2020, and in November that year the country fell back into civil war. This pits the federal government against the Tigrayan People's Liberation Front (TPLF), which controls the northern state of Tigray and until recently dominated federal politics. The TPLF fired the first shots on November 4th, when its fighters attacked a base housing federal troops – to pre-empt an imminent attack from federal forces, it said. In response, Abiy Ahmed, Ethiopia's prime minister, ordered air strikes and sent in ground forces. Hundreds, perhaps thousands, have been killed and many thousands displaced. In November 2020 the government declared that the TPLF had been "thoroughly defeated", and that most of its leaders had been killed or captured. But the fighting continued and Tigray's president, Debretsion Gebremichael, remained at large. He has accused the government of genocide. What is behind the conflict?

Tigrayans make up less than 10% of Ethiopia's population of around 110m, but the TPLF had dominated the coalition of ethnically based parties that ruled Ethiopia from 1991, when the TPLF and its allies overthrew the Marxist dictatorship of Mengistu Haile Mariam. (Eritrea broke away to become an independent country two years later.) The new Ethiopian regime won plaudits for unleashing economic growth, but notoriety for killing and torturing its opponents. Abiy came to power in 2018 on the back of mass protests by the country's biggest ethnic group, the Oromo. At first he freed political prisoners, welcomed back exiled political parties and promised free and fair elections. His peace deal with Eritrea, signed in 2018, won him the Nobel Peace Prize. Since then, however, he has arrested opposition leaders and his security forces have killed protesters. Many Ethiopians worry he is turning to authoritarian rule. When national elections, due originally in May 2019, were postponed indefinitely, ostensibly because of covid-19, the now-marginalised TPLF went ahead with its own regional ballot in September. Abiy's government deemed the polls illegal and slashed federal funding to Tigray, a move the TPLF said amounted to a declaration of war.

At the heart of the conflict are fundamental ideological differences about how Ethiopia should be organised. The TPLF and Oromo opposition leaders say Abiy is trying to dilute the ethnically based federalism that underpins the constitution. They accuse Abiy of trying to undermine the autonomy of Ethiopia's ten ethnically based states, and of centralising power. In contrast, Abiy's supporters say he is trying to reduce the divisive influence of ethnicity in politics and to build a stronger, more united Ethiopia. Abiy frames the conflict as an operation to restore "rule of law" over a faction that has become too accustomed to power. Other actors may have their own agendas. Eritrea's president, Issaias Afwerki, has no love for the TPLF. Eritrea lost a bitter war with a TPLF-led Ethiopia between 1998 and 2000 and witnesses say it has been helping Abiy's forces. Some Amhara militias, from the state to the south of Tigray, have also got involved in the fighting, perhaps seeing an opportunity to settle a long-running border dispute with their Tigrayan neighbours.

The conflict thus risks further inflaming ethnic tensions in Ethiopia. At worst, the result could be the balkanisation of Africa's second-most-populous country and the destabilising of an already fragile region. Across the country, Tigrayans have been harassed, Tigrayan soldiers in the federal army have been disarmed and Tigrayan civil servants told not to report to work. Tens of thousands of refugees have fled to neighbouring Sudan, many reporting atrocities carried out by Amhara militias and government soldiers. Prospects for peace look bleak. Some analysts have predicted a protracted guerrilla war. A convincing TPLF defeat, by contrast, might embolden Abiy to centralise more power. His supporters may hope to see a new federal arrangement, and perhaps a revised constitution, but it is unclear whether this would bring lasting peace.

In America, far-right terrorist plots outnumbered far-left ones in 2020

In the final days of America's 2020 presidential-election campaign, each side attempted to portray the other as a threat to domestic security. Joe Biden, the Democratic nominee, criticised President Donald Trump for refusing to disavow white supremacists, including the Proud Boys, a thuggish right-wing group, in the first presidential debate. Mr Trump responded: "Somebody's got to do something about Antifa and the left, because this is not a right-wing problem." The president insisted that Antifa, a loose collection of left-wing anti-fascism activists, was a domestic terrorist organisation, and far more troublesome than the far right.

A report released in October 2020 by the Transnational Threats Project at the Centre for Strategic and International Studies (CSIS), an American think-tank, found that far-right terrorism was in fact a much greater threat than far-left terrorism. The CSIS analysed 61 terrorist incidents reported between January 1st and August 31st and categorised them into four groups: religious, far-right, far-left and other. It drew on databases compiled by research groups and press releases from the FBI and the Department of Justice, cross-checked against criminal complaints and news reports. Hate crimes, protests, riots and civil unrest, including the disturbances after the killing of George Floyd by a policeman in Minneapolis, were excluded.

The CSIS found that far-right groups – including white supremacists, anti-government extremists and involuntary celibates, or "incels" – were behind 67% (ie, 41) of the terrorist attacks and foiled plots in 2020. Far-left groups were responsible for 20%, up from 8% in 2019. In all there were 71 attacks in 2019; except for a dip in 2018, the total number of attacks has been rising since 2014, driven mainly by an increase in right-wing terrorism.

The protests and unrest sparked by the Floyd killing were a popular target for right-wing groups. Vehicles were their weapon of choice. According to CSIS, 11 of the far-right terrorist incidents

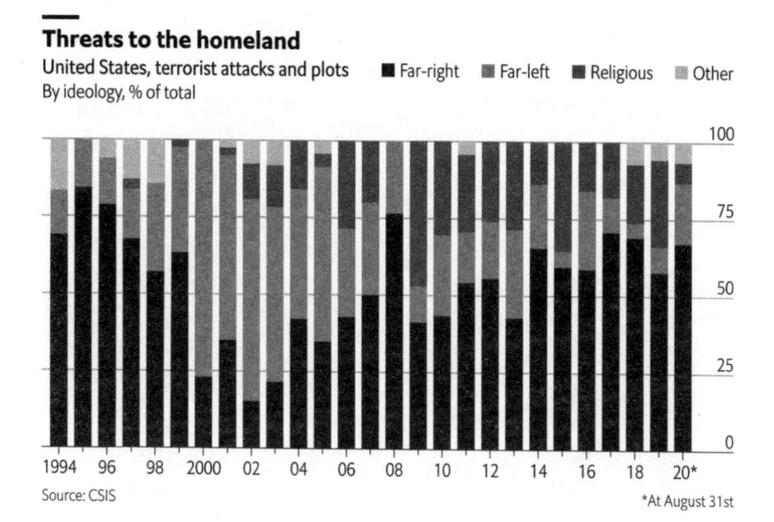

reported in the first eight months of 2020 were vehicle attacks, which require little expertise or resources but can still terrify, maim and kill. Only one such attack was carried out by the far-right in America between 2015 and 2019. Curiously, terrorist incidents were less lethal in 2020 than in previous years – only five resulted in deaths (excluding that of the perpetrator). In each of the previous five years between 22 and 66 people were killed.

The predominance of far-right terrorism is nothing new. Since 1994 more terrorist incidents have been associated with the far right than with all other groups combined. Over that period religious terrorism has caused more deaths – a grim distinction accounted for by the attacks of September 11th 2001 alone. But in 14 of the 26 years between 1994 and 2019, right-wing attacks were responsible for more than half of the fatalities, and in 2018 and 2019 for more than 90%. So far, American law enforcement has been slow to respond to far-right domestic terrorism. But there are signs that this might change. A report published in October 2020 by the US Department of Homeland Security labelled white supremacists as "the most persistent and lethal threat" to the country. The storming of the Capitol by right-wing sympathisers on January 6th 2021, in the wake of Mr Trump's defeat, has since raised tensions further. Far-left and far-right extremists see each other in action at protests and online. And so the threat of bloody escalation grows.

How are maritime boundaries determined?

A cannonball fired from the coast can hit a ship three nautical miles out to sea. In the 17th century, this was the basis for maritime boundaries, on the basis that countries could claim sovereignty as far as they could defend from the shore. Today, land is still the basis of claims over the sea, and with countries projecting their military and economic power farther across the waves, competing claims have left the oceans fraught with disputes: 39% of sea boundaries are yet to be agreed. However, this matters less than you might think.

Since 1982 oceans have been governed by the United Nations Convention on the Law of the Sea (UNCLOS). Countries calculate their maritime claims according to a simplified outline of their territory. For countries attached to a continental land mass, such as Italy, it roughly tracks the coast; for archipelagoes such as Indonesia, it skirts the farthest islands. Any water enclosed within this line, whether fresh or salty, is classed as "inland water". Countries have absolute authority here: no ship can pass through without permission, except on official sea lanes, and domestic laws apply as they do on land.

Beyond this zone, claims on the sea radiate out in rings of diminishing sovereignty. Countries can claim up to 12 nautical miles of "territorial sea": ships from other countries can sail through freely, but cannot fish, carry out military exercises or do scientific research. Beyond this is up to 12 nautical miles of buffer space, known as the "contiguous zone". Next comes the largest swathe, the "exclusive economic zone", extending up to 200 nautical miles from the coast. Here states have sole rights to drilling, fishing and mining. Where countries are close together and claims overlap, the boundary is normally set midway between them. Beyond these rings, countries can claim rights to their continental shelf (rock jutting out from their land below the surface of the sea) though not to the water above it. The rest of the oceans are international waters, where all countries have much more freedom.

UNCLOS was the result of decades of negotiation. The League of Nations, a forerunner to the UN, made a first attempt to broker a deal in 1930. In 1945 America claimed exclusive rights to its continental shelf and soon began drilling for oil in the Gulf of Mexico. Denmark, Norway and others followed suit in the North Sea, while Chile, Peru and Ecuador expanded their fishing zones to fend off pilfering by foreign ships. The scramble for offshore resources galvanised the international community, says Andreas Osthagen of the Fridtjof Nansen Institute, a research foundation near Oslo. UNCLOS expanded fishing and drilling zones, and formalised the right to make claims based on continental shelves. Although the cold-war superpowers, America and the Soviet Union, both benefited, newly independent former colonies also formed a powerful negotiating bloc.

The convention has now been ratified by 168 countries. The most notable absentee is America, whose treaty-shy Senate has resisted lobbying by the navy to sign it, though America expects China, which has ratified the convention, to abide by its rules in the South China Sea. The law is broadly accepted as customary, even by those who have not ratified it. It has channelled disputes into negotiation, arbitration or to the International Court of Justice (ICJ); where necessary it has let sleeping dogs lie. These mechanisms have not always worked. In 2016 an arbitral tribunal under UNCLOS upheld a case brought by the Philippines, one of several states in the South China Sea disputing China's territorial claims there. The tribunal rejected China's expansive claims, but there is no way to enforce the ruling. China has continued to build gun-laden islands around the contested Spratly archipelago. Most realists accept that China's claim, though invalidated by the international courts, is now a *fait accompli*, says Ben Saul of the University of Sydney.

Tensions in the South China Sea may be rising, but this dispute is "the fly in the ointment" for an otherwise successful treaty, says Michael Byers of the University of British Columbia. He suggests that most unsettled boundaries at sea go unnoticed and that even apparent hotspots can be misleading. NATO forays into the Arctic and the efforts by Canada, Denmark and Russia to claim the North Pole (based on continental-shelf arguments) mask co-operation among all three countries on surveying. Land-border conflicts, usually closer to civilians, rise to the top of the geopolitical agenda; tiffs at sea are simply not as significant, says Professor Byers. Yet such conflicts can also provide an opportunity for politicians to win favour with voters at home. During the Brexit negotiations, the British government cast fishing rights as an essential matter of national sovereignty, even though maritime borders between Britain and its neighbours were not in doubt, and the industry accounted for less than 0.1% of economic output. The symbolic value of fishing, says Philip Steinberg of Durham University, shows the power that the sea can exert over countries that tie their identities to the ocean.

A rift in democratic attitudes is opening up around the world

Is democracy on the rise or in decline? Proponents of either claim can often find supporting evidence in the headlines. Looking across the globe, however, academics generally agree that democracy is in a slump. One much-watched barometer is the World Values Survey (WVS), a poll published twice a decade. *The Economist* combined its results with data from the European Values Study to analyse trends in 98 countries from 1995 to 2020. Our analysis found that support for autocrats has indeed grown in most parts of the world – but this effect is weakest in healthy democracies, despite their recent flirtations with populism.

Since 1995 the WVS has asked people to rate several types of government as good or bad for their country. Among the options are "having a democratic political system", "having the army rule" and "having a strong leader who does not have to bother with parliament and elections". In the latest survey, only about a tenth of respondents were willing to describe democracy as a bad thing. However, nearly a quarter of them said that having the military in charge was a good thing, and more than two-fifths were in favour of strongmen who would ignore the outcomes of elections.

When analysing changes in these sentiments, we compared them with countries' actual levels of political freedom, as measured by the University of Gothenburg, in Sweden. On average, we found a big increase in support for despots in flawed democracies, but little change in places with lots of political freedom. For example, Mexico's approval for a strongman leader surged from 39% to 70% over the past two decades, whereas New Zealand's ebbed from 17% to 15%. Contrary to widespread fears about the death of democracy in the West, the share of people who think it is a bad system has actually fallen in the past decade in ten of the 15 most wealthy and free countries.

Using the WVS's individual-level data, we also built statistical models to predict the attitudes of a hypothetical person in a given

Vox populi

Strength of country's democracy v share agreeing with statement, %

From polls in 1995–98 and 2017–20*

1998 **2020**

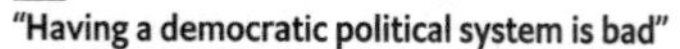

"Having a democratic political system is bad"

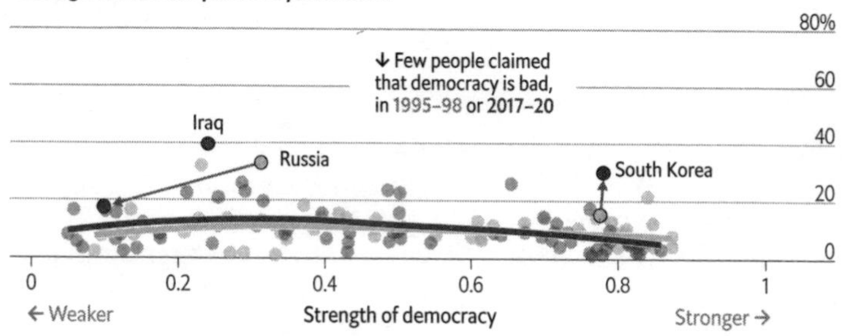

"Having the army rule is good"

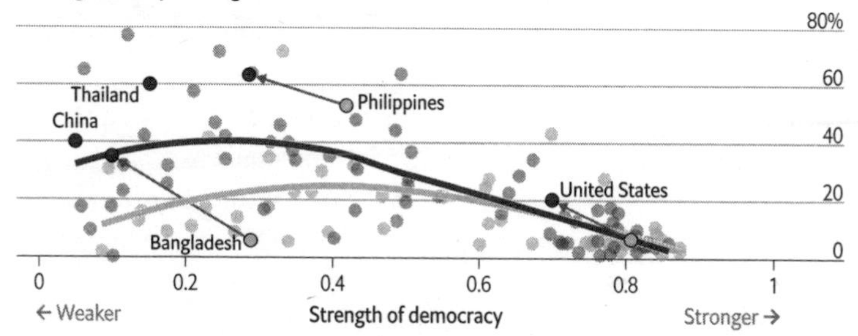

"Having a strong leader who does not have to bother with parliament and elections is good"

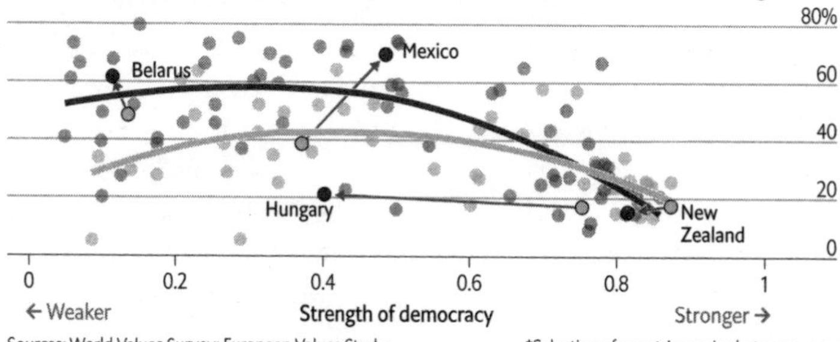

Sources: World Values Survey; European Values Study; V-Dem Institute, University of Gothenburg; *The Economist*

*Selection of countries varies between years

country and year. We then considered the opinions of a 40-year-old person with middling education, ideology and income, to test the impact of a change in a single characteristic on their overall beliefs. In highly democratic countries, our model predicted that the probability that such a person would favour a strongman rose from 29% to 33% between 1998 and 2020. But in countries with unhealthy democracies, the increase was far greater, rising from 44% to 62%. The growth was particularly high in Latin America, South-East Asia and former Soviet states.

Like previous analysts of the WVS, we found that young people tended to be the most susceptible to autocratic preferences. Right-wing and less-educated people also lean that way. But in countries where enthusiasm for despots is rising strongly, the growth seems to be similar across demographic groups. Whatever is driving people towards strongmen is affecting entire countries, not just cohorts within them.

Why Taiwan is not recognised on the international stage

There is an island 180km off the coast of China. Its democratically elected leaders say they run a country called the Republic of China. To the Communist government in Beijing the island is "Taiwan, China" or "Taiwan Province, China". International organisations, desperate not to offend either side, struggle to name it at all – many opt for the deliberately ambiguous "Chinese Taipei", after its capital city. To most it is just "Taiwan", a country that drew attention in 2020 for its exemplary handling of the covid-19 pandemic. Taiwan had been preparing for such a crisis since the SARS epidemic of 2003. To combat the spread of covid-19, it screened inbound air passengers and used national databases and big data to identify those most at risk of infection. Thanks to such measures and many others, Taiwan managed to suppress covid-19 without closing schools, restaurants and bars. Its success attracted sympathy in the West for Taiwan's efforts to secure observer status at the World Health Assembly, the annual decision-making forum of the World Health Organisation (WHO).

Taiwan is not a member of the United Nations, or any of its suborganisations, including the WHO. Between 2009 and 2016, when a China-friendly government held power in Taiwan, it attended World Health Assembly meetings as an observer. But since then it has been prevented from participating, at China's behest. America and New Zealand backed its bid to attend in 2020, but Taiwan decided to postpone its efforts to renew for observer status, citing a need to focus on the pandemic. Most of the assembly's members would not have dared to cross the government in Beijing, which insists that it alone represents China, and that China includes Taiwan. In 1971 the UN voted to recognise that government as China's sole representative.

Taiwan's woes stem from the unfinished business of the civil war that brought the Communist Party to power in China in 1949 and forced the deposed government of the Kuomintang (KMT) to flee to the island. The KMT continued to maintain that the Republic of China still existed, and refused to recognise the new People's Republic of China led by the Communists. It did accept that Taiwan was a province of China, but not of the People's Republic. In the 1990s, however, democracy began to take hold in Taiwan. This gave greater voice to politicians who saw the island as a country in its own right, with no lingering claims to the mainland. In 2000 an opposition party, the Democratic Progressive Party (DPP), won presidential elections for the first time. The DPP's assertion of Taiwan's separate identity enraged China, which responded by extending olive branches to the KMT – hence China's decision to allow Taiwan to attend World Health Assembly meetings during the years when the KMT was back in power on the island.

With the DPP in control again, China is unwilling to make any such concessions. Since 2016 it has been using carrots (large investments) and sticks (such as restrictions on visits by Chinese tourists) to persuade Taiwan's few remaining diplomatic allies to switch sides. Seven countries have done so, leaving only 15 that still recognise the Republic of China. The holdouts include eSwatini, Nicaragua and the Vatican. In January 2020 Tsai Ing-wen of the DPP won another four-year term as president, making a cross-strait thaw highly unlikely for the foreseeable future. Ms Tsai is widely viewed in Taiwan as relatively cautious in her handling of relations with China. But the government in Beijing remains suspicious of her party's pro-independence leanings. So China has continued to put pressure on other countries to freeze out Taiwan. Taiwan's absence from the World Health Assembly, despite its successes in fighting the pandemic, shows that this is working.

Faith in government declines when mobile-internet access arrives

The internet boom of the 1990s led to many predictions that access to information would unleash a wave of democratisation. More recently, views on the internet's impact have soured, as states have used it to spy on dissidents, spread misinformation and meddle in foreign elections. Opinions on this topic are abundant, but there is a shortage of hard data. No one knows whether the Arab spring would have occurred without the internet, or whether Russia's online efforts to boost President Donald Trump's campaign in 2016 had any effect. But scholars can sometimes find natural experiments to substitute for such counter-factual scenarios. A study by Sergei Guriev, Nikita Melnikov and Ekaterina Zhuravskaya, published in December 2020 in the *Quarterly Journal of Economics*, uses the growth of mobile broadband to reveal a link between internet access and scepticism of government.

Most of the 4.1bn people who are now online got connected after 2010. To measure how new users' views changed as a result, the authors combined two datasets. First, for each year between 2007 and 2018, they estimated the share of people in each of 2,232 regions (such as states or provinces), spread across 116 countries, who could access at least 3G-level mobile internet. Then they used surveys by Gallup, a pollster, to measure how faith in government, courts and elections changed during this period in each area.

In general, people's confidence in their leaders declined after getting 3G. But the size of this effect varied. It was smaller in countries that allow a free press than in ones where traditional media are muzzled, and bigger in countries with unfettered web browsing than in those that censor the internet. This implies that people are most likely to turn against their governments when they are exposed to online criticism that is not present offline. The decline in confidence was also larger in rural areas than in cities. A similar pattern emerged at the ballot box. Among 102 elections in 33 European countries, incumbent parties' vote-share fell by an average of 4.7 percentage points after 3G arrived. The biggest

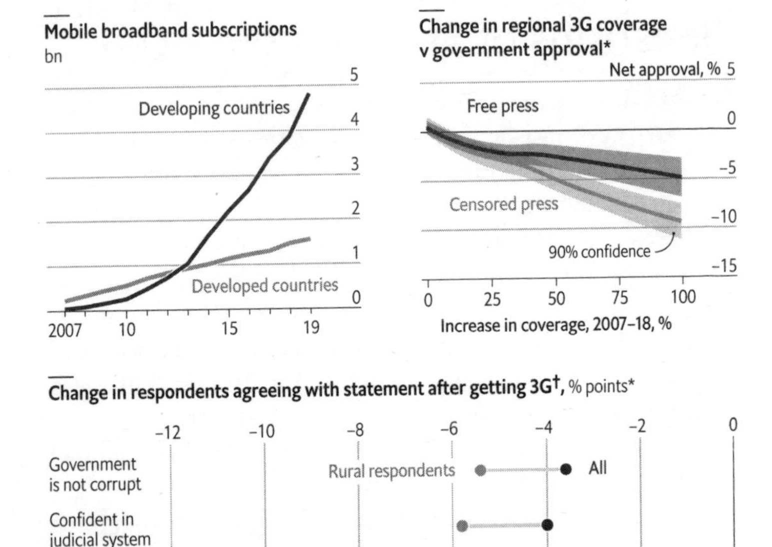

beneficiaries were parties classified as populist – though this may simply have been because they happened to be in opposition when voters turned against parties in power, rather than because of their ideology.

A central (and disconcerting) implication is that governments that censor offline media could maintain public trust better if they restricted the internet too. But effective digital censorship requires technical expertise that many regimes lack. In Belarus, where the government tries to control media both online and off, one opposition news channel on Telegram, an encrypted mobile app, has 2m subscribers – one-fifth of the country's population.

Questions of faith: religion and belief

Why Muslim scholars disagree over keeping dogs as pets

Among the many controversies that divide Muslim thinkers, the status of dogs is a minor one. But every now and then the debate is renewed. In August 2020 Egypt's grand mufti, Shawki Allam, weighed in on the side of man's best friend. Whereas some interpretations of Islam deem dogs impure, Mr Allam said that "It is possible to coexist with a dog and still worship God." Citing the Maliki school of Islam, he claims that every living animal is pure.

Conservative clerics are not rolling over, though. They have cited certain *hadith* (sayings of the Prophet Muhammad) to support their dogmatically anti-dog position. Angels will not enter a house if there is a dog present, says one. Another warns that a Muslim keeping a dog will lose out on some of the spiritual rewards from his good deeds. Yet there are caveats. Angels don't seem to mind if the dog is outside. And dogs used for herding, hunting, farming and guarding appear to be exempt from the rules. Perhaps the most contentious issue is a dog's saliva. "Cleanse your vase which the dog licked by washing it seven times and the first is with earth (soil)," goes another *hadith*. But Mr Allam countered that if you perform *wudu* (ablution) and a dog licks you, there is no need to rewash before praying. What about dog fur? Many scholars think it is clean and, therefore, petting is okay. But fierce debate ensues if the fur is wet.

The Koran itself says little about dogs. Scholars claim that the Prophet prayed among canines. A few years ago Osama al-Azhari, a religious adviser to Egypt's president, Abdel-Fattah al-Sisi, was photographed shaking the paw of a dog. Conservatives growled. But Mr Azhari pointed to the story of the Seven Sleepers – a group of young men who hid in a cave to escape religious persecution. According to the story, they took a 300-year nap while their dog, stretched out at the entrance of the cave, kept watch. (Conservatives note that the animal was not, however, actually in the cave.)

The resurgence of the debate in Egypt may have more to do with power than pups. Mr Sisi wants to assert more control over Islam via Dar al-Ifta, the state's Islamic advisory body led by Mr Allam, and at the expense of al-Azhar, a conservative seat of learning. Lately Dar al-Ifta has espoused relatively progressive views in order to win over the public. In general, Mr Allam has said that he wants to make it easier for Muslims to practise their faith.

Islam is hardly the only religion to feature confusing and contradictory texts – or to have mixed feelings about canines. Dogs are often used as a symbol of evil in the Bible. They are also portrayed negatively in the Talmud. To many Muslims, the whole debate is, you might say, a mastiff waste of time. In 2019 Iran tried to ban people from walking dogs in public. But the restriction had little bite, and many Iranians ignored it. More and more Egyptians appear to be keeping dogs as pets. Cats are also popular companions. Admired in Islam for their cleanliness, there is at least no debate about them.

How personal freedom varies across the Islamic world

In his writings, Mustafa Akyol, a Turkish-born author and columnist, has argued that there is no logical reason why Islam should not thrive in conditions of personal liberty. After all, in a much-debated verse, the Koran insists that "there is no compulsion in religion". But the reality of the Islamic world is drearily repressive, and has in some ways been getting worse, according to a report Mr Akyol published in August 2020 for the Cato Institute, a think-tank in Washington, DC. It looks mainly at 51 countries where Muslims make up at least a plurality of the population.

Only 60m of the world's 1.9bn Muslims live in countries where the general level of personal freedom is greater than the global mean, while more than 1.8bn live in places where liberty levels are clearly below average, he finds. To reach this conclusion he uses the Human Freedom Index, which the Cato Institute has developed in collaboration with two other think-tanks. It measures various entitlements, including freedom of movement, expression, identity, recourse to law and, of course, religion.

That total of 60m comes from adding up, to 30m, the smallish populations of the handful of Muslim countries which score fairly well on freedom – Albania, Bosnia, Burkina Faso and Kyrgyzstan – and then adding the 30m or so Muslims who live in Western democracies. At the other end of the scale are authoritarian states, like Egypt, Iran, Saudi Arabia and Sudan, and countries which have been convulsed by internal conflict: Iraq, Syria and Yemen. In between are the populous Asian giants with more Muslims than any other countries (Bangladesh, India, Indonesia and Pakistan) which are all nominal democracies, but all of which fall below the average score in the Human Freedom Index. It is striking that women's rights seem best observed in formerly communist countries which, although they may not be free in all respects, have at least retained secular law codes. These include Azerbaijan, Bosnia, Kazakhstan and Kyrgyzstan.

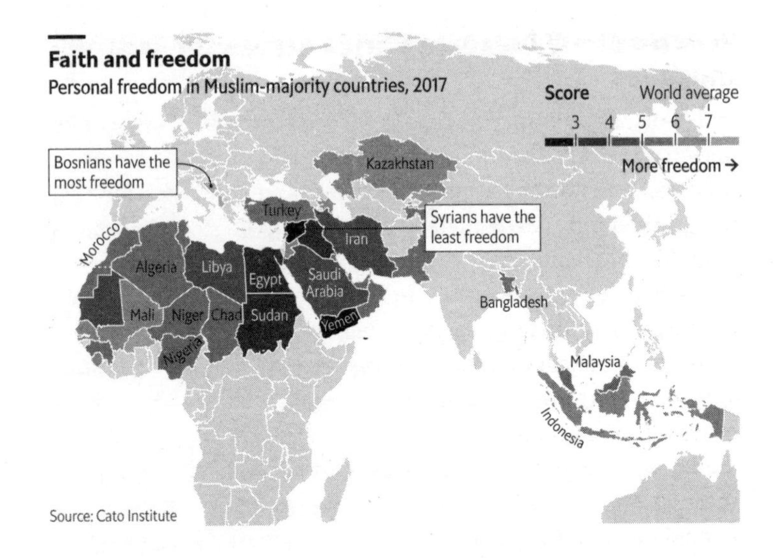

Mr Akyol laments the "dramatic deterioration" in all kinds of liberty over the past decade in Bahrain, Egypt, Syria and Turkey. He finds a more nuanced, and less depressing, picture when it comes to economic freedom: the right to offer goods and services without excessive state interference. On this measure, monarchies (Bahrain, Jordan, Qatar, the United Arab Emirates) do better than the global mean, whereas republics established in a burst of republican zeal, such as Algeria and Egypt, score dismally. Whether their regimes are traditional or secular, Muslim countries in the Middle East and north Africa are generally less free than those in say, central Asia or parts of west Africa. All this adds poignancy to the last article written by the Saudi journalist Jamal Khashoggi, entitled "What the Arab world needs most is free expression." It was published posthumously in the *Washington Post* in October 2018, shortly after his murder in Istanbul, in his own country's consulate.

Why so many Muslim leaders are building grand mosques

In the early days of Islam, mosques were modest affairs. The earliest ones had neither domes nor minarets. The Prophet Muhammad used his courtyard as a prayer hall. But Arab autocrats now see things differently. Many hope to leave a legacy in stone in keeping with their proclaimed grandeur.

Take Abdelaziz Bouteflika, Algeria's longtime president, finally ousted in 2019. Just as he was wheeled from office, his country beat Morocco in the competition to construct Africa's largest mosque, with a megalith costing $1bn that spans 40 hectares. It has all the charm of a vast Chinese airport with a traffic-control tower. Given that it was built by Chinese workers, that is almost what it is, minus a few archways. Sultan Qaboos of Oman, meanwhile, used to hold the record for hanging the world's largest chandelier in his state mosque. But in 2007 the al-Nahyans of Abu Dhabi (part of the United Arab Emirates) outdid him by hanging an even bigger one in the Sheikh Zayed mosque. Sinan Hassan, a Syrian architect, calls it "a Disneylandish propaganda tool".

Others are at it too. Egypt's President Abdel-Fattah al-Sisi adorned his new capital, to the east of Cairo, not just with a mega-mosque but with the Arab world's largest cathedral. Iran's ayatollahs in Tehran, for their part, are working on their own bid for the world's largest mosque, now half-built in cement and metal. "It's soulless, cold and brutalist with none of the ornamentation of tradition," says an Iranian architect. "It says we're huge – and bigger than you."

To be fair, grand mosque-building is a long-standing tradition. Within decades of the prophet's death, Muslim leaders had ordered up mighty religious buildings to cement and legitimise their rule. The Omayyads, a sybaritic early Muslim dynasty, built mosques like palaces, with golden mosaics and coloured cut marble. In the 15th century the Ottoman Turks began to fashion pencil-thin minarets to outdo the church spires in their newly conquered Christian lands. The worst tyrants often built the finest mosques. But at least they had style. "The most beautiful face the Earth has ever turned towards the sun," is how the writer Amin Maalouf described Samarkand, a city rebuilt in the 14th century by the Emperor Tamerlane, who was also known for erecting towers of skulls. If only today's tyrants had better taste.

Why American politicians are more pious than their constituents

America's constitution explicitly states that "no religious test shall ever be required as a qualification to any office". Yet Americans expect piety from their politicians. During the presidential election of 1800, William Linn, the first chaplain of the House of Representatives, argued that Thomas Jefferson was unqualified for office because of his "rejection of the Christian Religion". Jefferson, a staunch believer in the separation of church and state, rejected most orthodox teachings and, later in life, assembled his own Bible expunged of miracles and other supernatural "perversions". The election of such a man, Linn warned, would "destroy religion, introduce immorality and loosen all the bonds of society". (It did not.)

America's religious balance has changed since then. Today about 65% of Americans identify as Christian, down from 90% 50 years ago. The religiously unaffiliated – including atheists, agnostics and those without allegiance to a particular creed – are the fastest-growing group, accounting for about a quarter of the population. American politics does not reflect these changes, though. Except for Jefferson and perhaps Abraham Lincoln, all presidents have belonged to a church. The current president, Joe Biden, is Catholic, and nearly nine in ten lawmakers in Congress say they are Christian. Only two, Kyrsten Sinema, a senator from Arizona, and Jared Huffman, a congressman from California, claim no religious affiliation, identifying as "unaffiliated" and "humanist", respectively.

Demographics partly explains the discrepancy. Members of Congress are, on average, older than most Americans and belong to generations such as the baby-boomers, that are more likely to identify as Christian. Religiously unaffiliated Americans tend to be younger – nearly half of them are millennials, who are under-represented in Congress.

But even taking age into account, Christians are still over-represented on the Hill. There are other factors at play. In states such

The houses of the holy

United States Congress, by religious affiliation

Number of seats

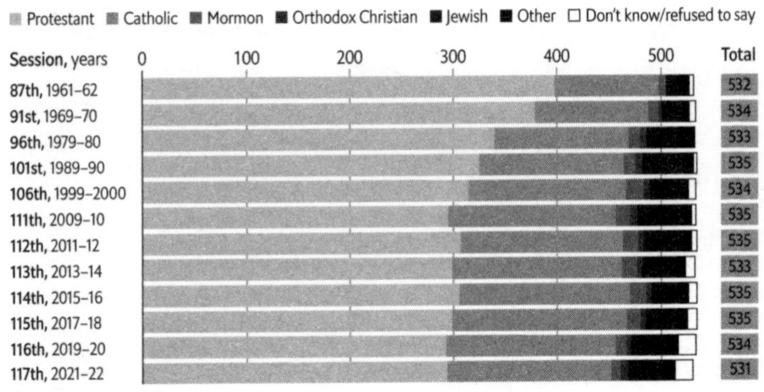

Source: Pew Research Centre

as Alabama, where 86% of adults identify as Christian, professing godliness is politically advantageous. And although suspicion of those without religious affiliation has abated since Linn's day, proclaiming yourself as such is still a political risk, says Ben Gaskins, a researcher at Lewis & Clark College in Oregon, who has studied how candidates project religiousness to establish trust with voters. "A lot of people still to this day believe that you can't truly be a good, moral person without a belief in God," says Mr Gaskins. A Pew Research Centre poll in 2014 found that no attribute – including inexperience, marital infidelity or age – made a voter less likely to support a presidential candidate than atheism. Donald Trump – a thrice-married political novice who once said he did not need to ask forgiveness from God – comfortably won the Christian vote in 2016. This suggests that claiming a religious identity, however tenuous, can cover a multitude of sins.

John Green, an adviser to Pew, says lawmakers often emerge from faith communities because "religious people of all kinds tend to be much more engaged in social groups... which is the stuff that politics is made out of". He also notes that certain communities have historically been flush with cash, giving them an edge in political representation. One example is Episcopalians, who have the highest household incomes among Christians. Episcopalians make up roughly 1% of Americans, yet they occupy 5% of seats in Congress and have sent 11 presidents to the White House, more than any other denomination.

Changes in America's religious make-up are being reflected in the halls of power. In the past decade and a half, Americans have sent their first Muslim, Hindu and Buddhist representatives to the Hill. In 2018 Mr Huffman co-founded the Congressional Freethought Caucus, a group that "protects the secular character" of American democracy and opposes discrimination against the religiously unaffiliated. It counts both believers and non-believers among its members. The unaffiliated are still under-represented in Congress, but they no longer face the kinds of accusations made by William Linn.

What do the French mean by "laïcité"?

Each time an act of jihadist terrorism is committed on French soil, the country is thrown anew into a global debate about *laïcité*, and whether it is the answer to the problem, or its source. The first article of the French constitution explicitly states that the republic shall be "indivisible, *laïque*, democratic and social". Difficult to translate, the word *laïque* refers to the French creed of *laïcité*, a form of secularism that is central to the country's history and identity, but much misunderstood elsewhere. It is neither a form of state atheism, nor the outlawing of religion. Rather *laïcité* enshrines in law the right to believe, or not to believe, while at the same time keeping religion out of public affairs. No French president, for instance, could ever be sworn in on a holy book. No French state school could hold a nativity play. No French marriage is legal if celebrated only in a place of worship.

Historically, French *laïcité* was the product of a struggle with the Catholic church. At the end of the 19th century, the republic's battle to wrest classrooms, the army and politics from the hands of the clergy was sometimes violent. Convents and religious schools were shut down by force; thousands of priests fled the country. "We have torn human conscience from the clutches of faith," declared René Viviani, a Socialist minister, in the National Assembly. This campaign resulted in a law in 1905 to entrench *laïcité* across France (with the exception of Alsace-Moselle, which enjoys a derogation under the Napoleonic Concordat of 1801, but that is another story). The point was to protect private religious belief, but also to keep public affairs free from the influence of any religion.

For the past 30 years, ever since three pupils were suspended in 1989 for wearing the headscarf to school in Creil, just north of Paris, the controversy over *laïcité* has shifted from the influence of Catholicism to the accommodation between the French state and Islam. Successive laws have (in 2004) banned from state schools and public institutions "conspicuous" religious symbols, including the Muslim headscarf (and crucifix) and then (in 2010) full face-coverings, including the niqab, from all public places. For liberal multiculturalists, such measures are a blatant infringement of the right to religious expression. Inside France, too, certain groups accuse successive governments of "weaponising" *laïcité* in a way that legitimises Islamophobia. They also point to a separate law on freedom of expression, dating back to 1881, which protects blasphemy and therefore legalises caricatures of the Prophet (and of Jesus, too). Emotions are heightened each time there is an official crackdown on militant Islamism, as often happens after terrorist attacks. The country is no stranger to anti-France boycotts and protests as a result.

For French secularists, however, who are to be found on both the political left and the right, this legal arsenal is a reaffirmation of the country's distinct model. It may be legal to insult a religion in France, but the law also forbids insulting or inciting hatred of any individual on the basis of that religion. Over the years, French governments have increasingly detected behind "soft" signs of conservative Islam an ideological effort to spread Islamism and test the resilience of the French model. President Emmanuel Macron has introduced a law against "Islamist separatism" that, among other things, bans home schooling on the ground that it can mask radicalised teaching. The outside world may look on in bemusement, but France is more likely than ever to continue to defend and reinforce *laïcité*.

Why there are more Christians in China than in France or Germany

Tearing down crosses from church spires may not sound the best way to win a promotion. But in Xi Jinping's China, it might do the trick. In 2014 Xia Baolong, the Communist Party chief in Zhejiang, a coastal province, oversaw a campaign to remove more than 1,500 crosses from places of worship in the province. Bibles were confiscated; pastors were locked up. It certainly did Mr Xia's career no harm. A longtime ally of the president, he was promoted first to a plum job in Beijing and then, in 2020, to a new assignment as head of the office overseeing Hong Kong and Macau affairs.

As for China's Christians, their numbers continue to grow. The government reckons that about 200m of China's 1.4bn people are religious. Although most practise traditional Chinese religions such as Taoism, and longer-standing foreign imports such as Buddhism, Protestant Christianity is probably the fastest-growing faith, with at least 38m adherents today (about 3% of the population), up from 22m a decade ago, according to the government's count. The true number is probably much higher: perhaps as many as 22m more Chinese Protestants worship in unregistered "underground" churches, according to a study by researchers at the University of Notre Dame. As China also has 10m-12m Catholics, that means there are more Christians in China today than in France (38m) or Germany (43m). Combined, Christians and the country's estimated 23m Muslims may now outnumber the membership of the Communist Party (92m). An unknown number of party members go to church as well as local committee meetings.

But as they become more numerous, the country's faithful face ever-stricter oversight from the state. China's constitution nominally guarantees freedom of religious belief. But since Mr Xi came to power in 2012, the government has tightened its control of religious groups in an effort to eliminate possible sources of dissent or secession. Government-approved versions of traditional faiths have been promoted, and those seen as foreign and threatening

Spreading the word
China

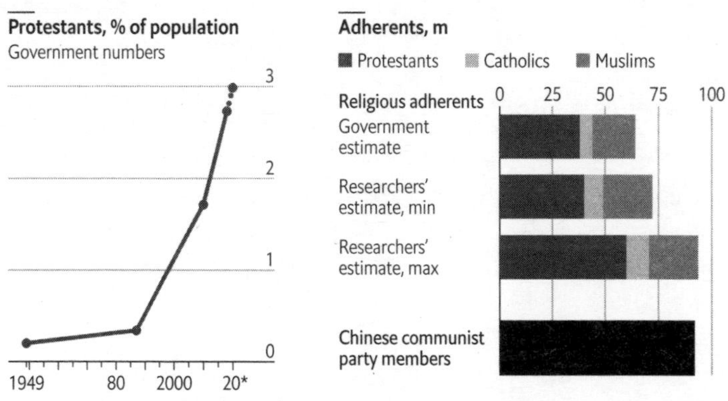

Sources: "Official Protestantism in China", K. Koesel, Yizhi Hu and J. Pine, *Review of Religion and Chinese Society*, April 2019; "China's old churches", A. R. Sweeten, *Studies in the History of Christianity in East Asia*, December 2019; "Anti-Muslim bias in the Chinese labour market", Yue Hou, Chuyu Liu and C. Crabtree, *Journal of Comparative Economics*, June 2020 *Forecast

have been repressed. For hundreds of thousands of Uyghur Muslims in the western province of Xinjiang, this has meant detention, "re-education" and forced secularisation. For many Christians it has meant toppled crosses, closed churches and, in some cases, prison. In 2019 Wang Yi, the pastor of an underground church in Chengdu, capital of the south-western province of Sichuan, was sentenced to nine years in prison for "inciting subversion", a charge normally reserved for political dissidents.

Preachers face harsh surveillance. Whereas the clergy at state-sanctioned churches are told to lecture on party-favoured topics, such as blending Christianity with secular Chinese culture, underground pastors often touch on taboo subjects, such as the existence of demons and miracles, or even worse, the importance of proselytising to friends and co-workers. Yet research suggests the government may need to continue to tolerate some more adventurous sermons in state-sanctioned churches, or risk losing their congregants elsewhere. In 2017 and 2018 Harris Doshay, a doctoral student at Princeton University, attended and analysed the sermons delivered by ministers of the Three-Self Patriotic Movement, the state-controlled Protestant church. Mr Doshay found that congregants showed their preference by "voting with their eyelids": if the sermons stayed within party lines, many, consistently and sometimes rather demonstratively, decided to nap.

Which countries still have blasphemy laws?

All over the world, people are being attacked, imprisoned, tortured and occasionally executed on the grounds that they have said or done something which offends somebody else's religious beliefs. Usually it is the state that punishes such acts of blasphemy, though often violent mobs make it their business, and the two forms of retribution cannot fully be separated. That is the gloomy picture painted by a report on blasphemy laws issued in 2020 by the United States Commission on International Religious Freedom, a bipartisan body charged by Congress and the president with monitoring liberty of conscience across the world.

The study, which analysed 674 cases of state enforcement of criminal blasphemy laws between 2014 and 2018, found that the great majority were carried out by just ten countries: in descending order, Pakistan, Iran, Russia, India, Egypt, Indonesia, Yemen, Bangladesh, Saudi Arabia and Kuwait. Of these, all except Russia and India are countries where Islam (or at least monotheism, in Indonesia) is entrenched in the political system and followed by the great majority of people. These two outliers are a reminder that the harsh use of blasphemy laws is not confined to the Islamic world. In theory the Russian and Indian constitutions are religiously neutral, but in practice each has a preponderant faith (respectively Orthodox Christianity and Hinduism), and both religious minorities and secularists complain of discrimination. Russia has arrested scores of Jehovah's Witnesses, for example, as well as followers of mystical Islamic movements.

Pakistan alone accounted for a quarter of the cases of state enforcement and nearly half the incidents of mob violence against perceived blasphemers. Vengeful crowds were also on the march in Bangladesh, Egypt and Nigeria. Sometimes state authorities use the anger of the mob as an excuse for their own acts of repression, the report adds. For example, Mohamed Ould Abdel Aziz, Mauritania's former president, justified the detention of a sacrilegious blogger between 2014 and 2019 on the grounds that "millions of people" wanted to see him executed and his safety was at risk.

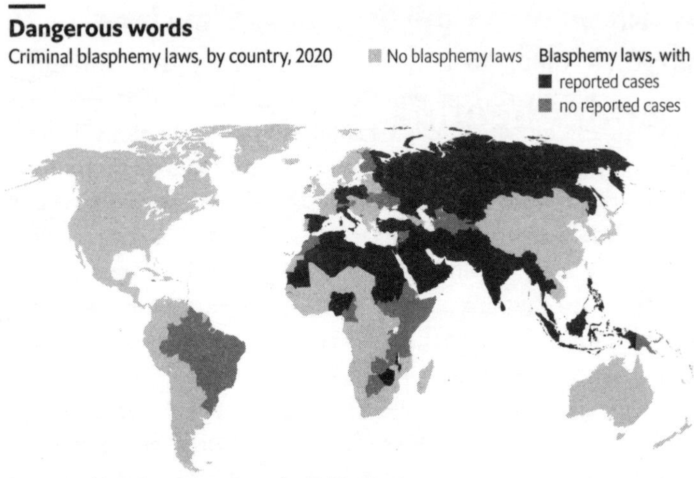

The report was warmly received in the American Congress, where the House of Representatives voted almost unanimously in December 2020 to put the repeal of blasphemy laws higher on the foreign-policy agenda. Lawmakers urged the White House and State Department to "make the repeal of blasphemy, heresy and apostasy laws a priority" in relations between the United States and religiously repressive countries. Jaded diplomats will retort that religious freedom is but one of many topics that America needs to broach with the ten countries named as the worst offenders.

Gender agenda: sex, marriage and equality

The sad truth about Vegas weddings

"Get 'married' to your best friend for life, your longtime partner, your cellphone, anything really!" reads the brochure for The Little Vegas Chapel. Its Pretend Wedding package ($199) promises all the thrill of a wedding, without the lifetime obligation. Many Las Vegas chapels now offer a range of non-binding ceremonies to pad out thinning wedding schedules. Vow-renewals account for much of their business, alongside friendship affirmations and business-partner commitments.

The self-proclaimed "wedding capital of the world" is suffering from millennials postponing marriage, perhaps indefinitely. About 40% fewer licences were issued to couples in their 20s and 30s in Clark County in 2018 than a decade earlier. Overall, marriage licences issued were down by 42% from their peak in 2004, which meant that the local economy missed out on $1bn of annual revenue. A Vegas wedding was supposed to be a counter-cultural choice, but it turns out to have been tied to the very traditional habits it was supposed to subvert.

Las Vegas's Hollywood image as the backdrop for impulsive romantic unions is divorced from reality. Its marriage-licence bureau's 24-hour service, which started in 1979, stopped in 2006. Instead, it now closes at midnight – not to deter any regrettable 4am decisions, but for lack of trade. By the time the office changed the rules, only 4% of its customers were pitching up after midnight. Ron DeCar, an Elvis impersonator and owner of the Viva Las Vegas Wedding Chapel, has seen customer numbers, and hence his takings, fall every year for over a decade. He has been inventive in attempts to arrest the decline, offering 14 variations on his standard Elvis package. For its Blue Hawaii wedding package ($795), the chapel is filled with palm trees and an atmospheric "ocean mist". A dancer dressed as a hula girl, or Priscilla Presley, as preferred, is there to assist Mr DeCar as officiating Elvis. For the Pink Caddy Luxury Option ($1,616), the bride is driven to the altar in a 1964 convertible Cadillac through candlelit dry ice. Live-streaming means that guests can watch the whole thing remotely.

For all the creativity, this chapel, like the others, faces a shaky future. One of the most venerable, A Little White Wedding Chapel, the venue for about 800,000 ten-minute weddings, failed to find a buyer after six months. Even fame as the chapel of choice for Britney Spears, Frank Sinatra and Michael Jordan did not help.

To drum up trade, the clerk's office began operating a pop-up licence booth at the local airport in 2020. Its aim is to make marrying more expedient – and perhaps tempt newly arrived couples. Christine Crews of the airport's public affairs department enthuses that between the booth, flower-vending machines, jewellers and liquor stores, the airport now offers all you need for an impromptu proposal. But however slick the set-up, the most important ingredient – the willing couple themselves – is becoming increasingly elusive.

Why it's so hard to predict the peak of world population

The world's population may never grow as large as had previously been assumed. In a paper published in July 2020, researchers at the Institute of Health Metrics and Evaluation (IHME) at the University of Washington predict that the global population will top out in 2064 and then fall steadily. The latest estimates by the UN's Population Division, by contrast, reckon it will continue to grow until at least 2100. As a result, the IHME estimates a total population of 8.9bn in 2100; the UN places the number at about 10.9bn.

The huge discrepancy is largely accounted for by differing views on two issues. First, the IHME study's central scenario assumes that improvements in access to education and contraceptives in sub-Saharan Africa – and a concomitant fall in fertility – will result in a population there of just under 3.1bn in 2100, compared with 3.8bn in the UN study. Accounting for mortality, this means 890m fewer African births on a cumulative basis in the remainder of the century. However, even the IHME's conservative projections still have sub-Saharan Africa as the only part of the world with a growing population by the end of the century.

Other demographers have also argued that better education could cause the global population to peak before the end of the century, because girls who go to school end up having fewer children. In 2019 Wolfgang Lutz, a demographer at the International Institute for Applied Systems Analysis in Austria, estimated that if recent progress in education in sub-Saharan Africa was maintained, the world's population would peak at 9.4bn in 2075 and decline to 8.9bn by 2100 (as the IHME also estimates). Brisker progress would imply an earlier peak, at 9bn, and a fall to 7bn by the end of the century.

The second reason for the discrepancy between the IHME and UN estimates is that the former is more conservative about what will happen to populations when fertility rates fall below the replacement level of 2.1 births per woman. The UN assumes that in

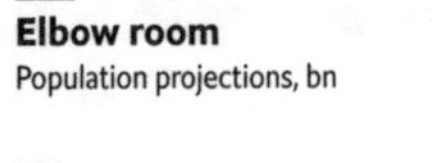

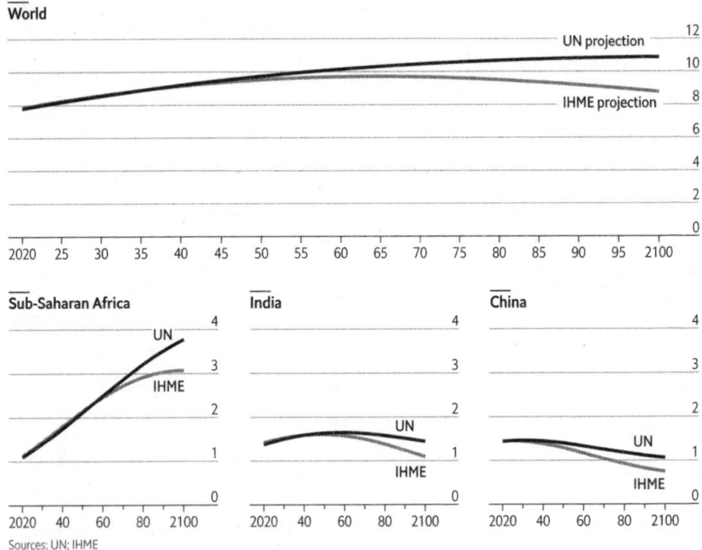

many countries with low rates, such as Taiwan, where the fertility rate is just over 1, they will increase again – not all the way to 2.1, but to around 1.75. The IHME believes that 1.4 is a more likely outcome. The difference yields a substantial gap in population projections.

Some of the IHME's projections are eye-popping. South Korea, a country of 52m people, and one of the greatest economic success stories of the past 50 years, is expected to have fewer than 27m in 2100. Spain is likely to lose more than half of its 2017 population by 2100 and the number of Bulgarians may fall from just over 7m to 2.6m. In total, 55 countries will experience a population decline of at least 25%. In 23 of those, the fall will be greater than 50%. Some rich countries, such as Australia and New Zealand, will continue to have growing populations, owing largely to immigration. Meanwhile, the populations of India and China, currently the world's two biggest, will fall from their current levels of 1.4bn and 1.6bn to 1.1bn and 730m respectively.

Should these outcomes come to pass, they would be broadly positive for the environment and the fight against climate change, at least relative to the UN's expectations. Fewer people, other things being equal, ought to mean lower carbon emissions and lower demand for food and water – although rising living standards may have a countervailing effect. Having fewer adults of working age would also mean fewer taxpayers. Without increases in productivity, or longer working lives, that would also mean slower economic growth. Social security and welfare systems would come under ever-greater stress. And labour would become scarcer – and so perhaps better paid.

What is being done to tackle "period poverty" in the West?

For the vast majority of menstruating women in rich countries, tampons and other feminine-hygiene products are both affordable and, on the whole, adequate. But some still end up soaking up their blood with poor substitutes such as newspapers, toilet roll or rags, all of which may lead to infection. This aspect of destitution, known as period poverty, has been the subject of a wave of activism in recent years. A dearth of data makes it difficult to grasp the scale of the problem in the West. But research shows that poor menstrual hygiene is not limited to poor countries. Two points stand out.

First, some women and girls struggle to pay for pads and tampons. According to a study in 2019 of 58 girls at a secondary school predominantly attended by poorer students in St Louis, Missouri, nearly half had been unable to afford period products when they needed them at least once during the previous year. Almost two-thirds had used products provided by their school. Nearly two-thirds of low-income women surveyed in the same city in a study the previous year reported being unable to purchase pads and tampons because they were out of pocket. Several admitted to stealing pads or tampons out of desperation. If a woman is worried about leakage or nasty smells, she is less likely to turn up at work or concentrate at school. Women who have lost income or jobs as a result of the pandemic may find the expense of sanitary products even harder to bear.

Second, there is an enduring awkwardness when it comes to speaking about periods, and menstrual conditions in general. Nearly half of girls in Britain felt embarrassed by their monthly cycle, according to an online survey of 1,000 girls conducted in 2017 by Plan International UK, a children's charity. This figure may not be representative (opt-in polls will probably be ignored by those who are not interested in the subject or unwilling to talk about it), but it indicates that a sense of shame about periods lingers. Boosting education about menstruation and medical conditions related to it, such as endometriosis, would help. England made learning about menstruation, hygiene and what counts as a healthy period compulsory for all students in September 2020, and in November that year Scotland became the first country in the world to pass a law that will offer free pads and tampons to all women via public buildings, schools and universities, and through a voucher-like scheme.

Other governments may balk at the cost of implementing a similar programme, particularly because most women have no trouble buying a box of tampons costing £2.20 ($2.98) or so on in Britain. A cheaper alternative, chosen by several countries and some American states, is to supply menstrual products to schools and hospitals. These are places where people are more likely to be caught short. This solution helps girls to manage puberty and avoids spending money on women who are happy to buy their own sanitary items. These policies also win rare bipartisan support in America, notes Jennifer Weiss-Wolf, an activist and a fellow at the Brennan Centre for Justice, a non-partisan think-tank. What about the "tampon tax", the sales tax levied on sanitary products? Britain abolished value-added tax on such goods from January 2021. A handful of American states and 15 countries, including Kenya and Nicaragua, have also scrapped taxes on sanitary wares.

Ensuring that all women have a hygienic and dignified period requires thinking about public-health and gender issues more broadly. Female-friendly toilets and menstrual education need to be part of the package too, says Marni Sommer of Columbia University's Mailman School of Public Health. The needs of menstruating trans and non-binary people, who may feel uncomfortable using single-sex toilets, raise further questions. But increasingly, lawmakers are beginning to agree with activists: periods should not stop a woman from going about her daily life.

Why east and west German women still work vastly different hours

Researchers are often greatly excited by "natural experiments", events that end up separating two groups of people, allowing wonks to compare their subsequent behaviour. Much like the study of twins adopted into different households, the post-war division and eventual reunification of Germany could be seen as such an experiment. A report by the German Institute for Economic Research on working mothers, published in October 2020 ahead of the 30th anniversary of reunification, reveals the interplay between policy and attitudes that influences the decision to work.

When the German Democratic Republic (GDR) in the east united with the Federal Republic of Germany (FRG) in 1990, the mothers of young children led very different lives. Life expectancy and incomes were much lower in the east, but communism did at least seem to lead to greater gender equality in labour-market outcomes. Encouraged by state policies and party ideology, mothers were almost as likely to work as fathers, and most worked full-time. In the west, where state and church encouraged mums to stay at home, less than half were in paid employment, and most of those worked part-time.

Three decades on, how has the picture changed? Two things stand out. First, behaviour has changed drastically since unification: the share of eastern women with young children working full-time fell from over half in 1990 to just under a third in 2018 (see the left-hand chart). More women across Germany now work part-time. Second, east-west differences still exist. The share of eastern mums in full-time work is more than double that in the west. As a result, whereas women in the east earn 7% less than men, the gap in the west is 22%. The report argues that policy and attitudes together explain these trends.

Policy seems to play a powerful role in explaining the collapse in full-time employment in the east. Despite some recent changes, the policies of unified Germany, like those of the FRG, still assume

After the wall
Germany

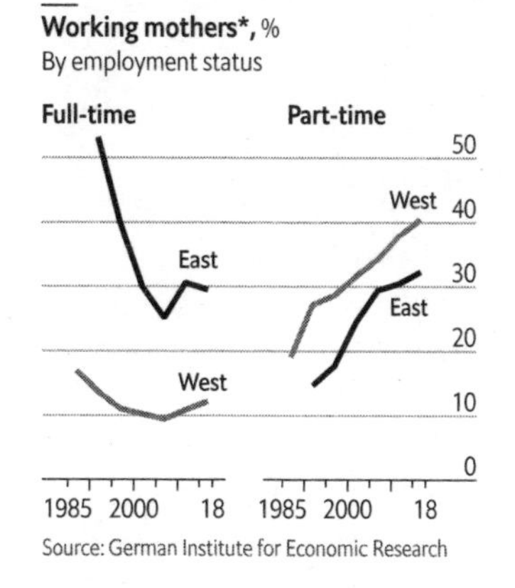

"A mother who works full-time can usually have just as meaningful a relationship with her toddler as one who does not work"
2016, % of women agreeing

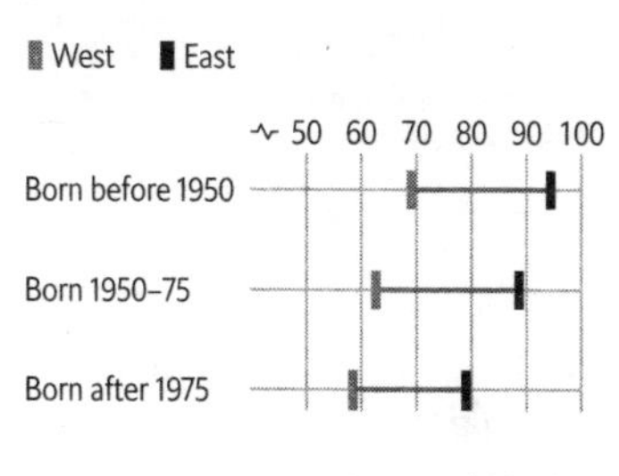

*Youngest child under 11

that women are wives and mothers first. Joint taxation of married couples, free co-insurance for spouses and tax breaks for "mini jobs", or low-hours contracts, probably did little to encourage women in the west to up their hours, and put those in the east off full-time work.

Attitudes, meanwhile, may help explain part of the lasting hours gap between east and west: 30 years after unification, eastern women are still more likely to approve of full-time working mums (see the right-hand chart). This chimes with earlier findings that east Germans are more likely to have an egalitarian view of the roles of the sexes. Attitudes have also changed over time, though. Strikingly, women born after 1975 in both the east and west are more likely to disapprove of mothers in full-time work than older ones, putting paid to the idea that younger women are keener on work. Perhaps women's views are shaped by the policies they face. Katharina Wrohlich, one of the report's authors, also suspects that the shift marks a rejection by younger women of both the dual-earner model of the GDR and the single-earner model of the FRG. "Instead, the younger generation seems to be aspiring to the one-and-a-half jobs model," she says – a preference that policy may need to respond to in turn.

The unification "experiment" hardly took place in laboratory conditions. Many women migrated from east to west. The regions differ in many other respects – incomes per head are lower in the east, for instance – that also affect the number of hours women work. But the episode still says something about the power of policy and the endurance of attitudes, long after walls are torn down.

Why the lot of female executives is improving

Wall Street's glass ceiling cracked at last on March 1st 2021, as Jane Fraser took charge of Citigroup, becoming the first woman to head a big American bank. That cracking sound has also been echoing across the rest of America Inc. In 2020 Carol Tomé became boss of UPS, a package-delivery giant. In January 2021, Rosalind Brewer became only the third black woman ever to run a Fortune 500 company (Walgreens Boots Alliance, a pharmacy chain). A month later Thasunda Brown Duckett was picked to run TIAA, a big pension fund.

Yet despite progress for women in the workplace, America still has a long way to go, according to *The Economist*'s glass-ceiling index, which ranks conditions for working women across 29 countries. As usual, Nordic countries performed best overall in our 2021 ranking, with Sweden, Iceland, Finland and Norway taking the top four spots. At the bottom is South Korea, which scored just 25 out of 100 on our index, less than half the average for the OECD, a club of mostly wealthy countries.

The glass-ceiling index

Environment for working women, 2020 or latest, 100=best

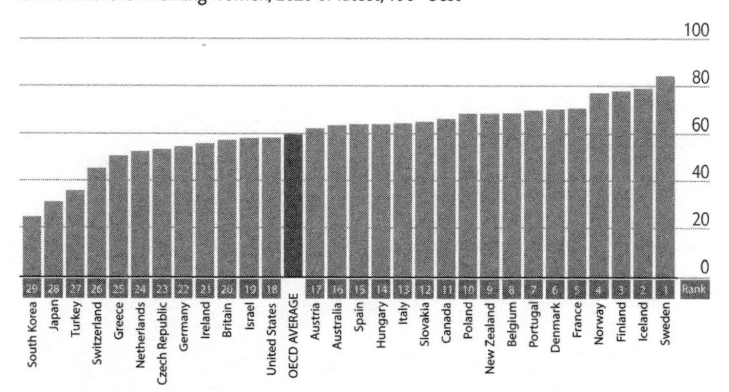

Sources: European Institute for Gender Equality; Eurostat; MSCI ESG Research; GMAC; ILO; Inter-Parliamentary Union; OECD; national sources; *The Economist*

America received poor marks on parental leave and political representation. But it has a high share of women in management (41%) and on company boards (28%). In both cases America outpaces egalitarian Germany, which in January 2021 enacted a quota for female board members (and where the shares for management and board-rooms are 29% and 25%, respectively). On average, just one in three managerial positions across the OECD's 37 members is occupied by a woman. A study by SIA Partners, a consultancy, found that in Britain bias against women in senior corporate hiring remains systemic, with job ads for high-ranking positions using more "masculine" words that make them less appealing to women.

At least signs of progress can be seen, even in traditional laggards like Japan. Mori Yoshiro had to resign as chief of the Tokyo Olympics in February 2021 after he complained that women talked too much in meetings. He was replaced – by a woman.

How to shrink America's gender pay gap

There is no shortage of cringe-worthy questions an eager job applicant might encounter during an interview. "What is your biggest weakness?" "Who are your most profound influences?" "What makes you a better candidate than the person we interviewed just before you?" But in many states and cities in America, the prickliest question of all – "How much did you make in your last job?" – is now illegal.

In 2016 Massachusetts became the first state to prohibit employers from asking job applicants about pay in previous jobs. Since then such salary history bans (SHBs) have been passed by 17 other states and by 21 municipalities, according to *HR Dive*, a human-resources trade publication. The goal is to make the application process fairer. Allowing employers to know candidates' salary histories, policymakers argue, encourages discrimination against those who have been out of the labour force – perhaps caring for children – or whose past positions do not adequately reflect their skills and qualifications. Advocates say this form of discrimination is felt most acutely by women, and that SHBs can therefore help level the playing field.

A working paper published in April 2020 by Benjamin Hansen of the University of Oregon and Drew McNichols of the University of California, San Diego, suggests that they do. Using earnings data from the US Census Bureau's Current Population Survey for the years 2006-19, Messrs Hansen and McNichols calculated women's earnings, relative to men's, by state, age and industry. To estimate what would have happened to wages in the absence of SHBs, the authors created "synthetic" control groups of states, similar in every way to the affected states except for the new laws.

The researchers found that, after California's ban was introduced in 2018, the gender pay gap for women over 35 narrowed by 2.3 percentage points (around 9.5%). For married women with children over the age of five, the gender pay gap shrank by 4.7 points. Many women in this group may be returning to work after a spell looking after children full-time; without an SHB, they would have suffered a bigger pay penalty. The authors repeated the experiment for other states with SHBs and found a similar pattern. The effect was particularly strong among older workers and those in male-dominated industries.

These findings suggest that concealing salary histories may help narrow the gender pay gap. But making everyone's pay more transparent is also a potent measure. An analysis of faculty salaries at public universities in Canada in 2019 found that pay-transparency laws reduced the gap there by two percentage points, or roughly 30%. Why do full secrecy and full transparency about pay have similar effects? Perhaps it is because both measures ensure that employers and employees have access to the same information. When both sides have no data, or all of it, it becomes harder to discriminate against women.

Why women are less likely than men to die from covid-19

Men are around 1.7 times more likely than women to die from covid-19, according to a paper by researchers at Yale University, published in January 2021 in the journal *Science*. Men older than 30 have a significantly greater mortality risk. That is striking, but perhaps should not come as a surprise. Human lifespans have climbed steadily in recent decades, but wide disparities remain between women and men. Globally, women live nearly five years longer than men, up from three years longer in 1950. The reasons for that gap are both biological and environmental – and help to explain why more men are dying from covid-19.

For one thing, men smoke and drink more. In Russia, where male drinkers outnumber female ones by four to one, alcohol was responsible for about three-quarters of all deaths among working-age men during the 1990s, widening the gender gap in life expectancy to more than 12 years. It is now down to ten years, the same as in neighbouring Belarus and Ukraine, thanks in part to a big reduction in drinking. Boozing contributes to conditions such as heart disease and liver cirrhosis. These sorts of comorbidities make it harder for men's bodies to fight off a disease such as covid-19.

Aggressive and risky behaviour also plays a role. Men are more likely to die violently, in car crashes or in other accidents. In El Salvador, one of the few countries where the life-expectancy gap between the genders has increased since 2000, gang violence is partly to blame. Men are also less likely to seek medical help than women. This may explain the large disparities in countries with high rates of HIV/AIDS and tuberculosis, such as eSwatini, Mozambique and Namibia. Women in sub-Saharan Africa are more likely than men to get diagnosed, start treatment earlier and stick to it. The result is that women account for 59% of HIV infections, but only 47% of HIV-related adult deaths. When it comes to covid-19, this combination of risk and recklessness may result in more men

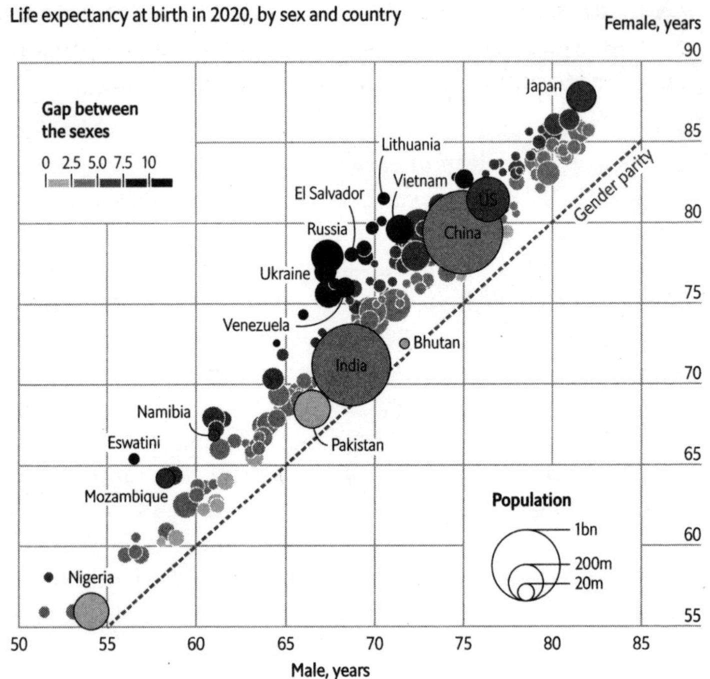

exposing themselves to the virus and then leaving it too late to seek treatment.

There are biological differences between the sexes, too. Women have stronger immune systems, thanks in part to oestrogen, a hormone that stimulates the body's defence response. Double copies of X chromosomes also protect women from genetic mutations and are responsible for immune-related genes. And the authors of the *Science* paper note that detrimental effects of ageing on the immune system typically occur five to six years earlier in men than in women, making it harder for men to fight off diseases as they get older. Men should not take these factors as a death sentence. They can improve their chances of beating covid-19, and extend their expected lifespans, by living more healthily, taking fewer risks and seeking medical care. But women will probably continue to outlive men – and to survive covid-19 in greater numbers.

Why America's demographics are looking more European

For generations, demographers considered America a standout. Lots of immigration and relatively high fertility rates increased its population faster – and kept it more youthful – than its rich-country peers. Americans were also exceptionally mobile within their borders. Over many generations they proved much readier than Europeans, for example, to flit between cities (or states) in search of a new job or lifestyle. That dynamism helped to produce a flexible labour force and lively economy. Now the exception is waning. Several big states have lost their demographic oomph. In December 2020 the US Census published new population estimates. For those keen on growth, they offer mostly grim reading. California's population has plateaued and may, for the first time, be declining. Illinois, which has shed over 250,000 residents in a decade, has shrunk for seven successive years. And in the year to July 2020, thus counting in little pandemic effect, New York endured more shrinkage than any state: it lost 126,000, or 0.65%, of its people. Some states, mostly in the South, are growing fast, but not enough to lift the national rate.

Overall, America's population is barely inching up by historical standards. In the year to July 2020 it grew by 0.35% (or 1.2m) to 329m. No year since 1900 has seen such a miserly gain. Even in the dark days after the first world war, as the Spanish flu raged, growth was faster. William Frey of the Brookings Institution calculates expansion in the decade to July 2020 at just 6.6%. If his sums are right, that is the lowest decadal gain since 1790. Evidence has also piled up to show Americans becoming much less mobile. Mr Frey notes a smaller share of them moved house in the year to March 2020 than at any time since reliable figures were first gathered in 1947. Just 9.3% of the population moved, barely half the figure in the 1980s, part of a steady decline over decades. Many factors contribute to that, including the high cost of housing that makes it difficult for younger Americans to move.

Kyle Mangum of the Federal Reserve Bank of Philadelphia published a paper in 2020 analysing why people move less frequently than their parents did, saying many factors – especially the absence of new, fast-growing cities and more similarity between various labour markets – mean that "this nation of pioneers has parked its wagons". He also notes how technology, such as air-conditioning, previously did much to open up territory for settlement. More recent technology, notably the internet, may instead have made it less necessary to move to find work.

Various changes reinforce each other. Less immigration, for example, has several effects. The 2010s saw the smallest expansion of the foreign-born population in any decade since the 1970s. Far fewer immigrants are flowing to big cities such as Los Angeles, New York and Chicago, helping to explain why they are not growing. Lower immigration hits domestic mobility, because recent immigrants are among the readiest to move for work. It may also bring down fertility rates. The average American woman is now expected to have 1.7 children in her lifetime, the lowest level in decades. That is below France's rate (at 1.9), on par with Britain's, and only slightly ahead of Canada's (1.5).

The result is more demographic stagnation. Mr Mangum sees a long-term reversion to the mean as America becomes less of an exception among rich countries. More recent influences accentuate that. The policies of Donald Trump sharply cut inflows of migrants. Joe Biden's administration is reversing some of them but probably not all, or at least not quickly. The pandemic has its own effects. Two academics at UC Berkeley, Joshua Goldstein and Ronald Lee, suggest deaths from covid-19 will cut average life expectancy by more than one year. The pandemic and its associated economic slump may also have caused a baby bust. Two researchers, Melissa Kearney at University of Maryland and Phillip Levine at Wellesley College, estimate that there will be 300,000 fewer births than otherwise expected in 2021 (there were 3.7m in 2019).

Post-pandemic, some of this could be reversed. Those who put off having children may cause a brief baby boom in 2022 or

2023. Eventual reopening of borders should see immigration tick up again. Even so, slower population growth will "continue in the coming years" says Joseph Chamie, a demographer in Oregon, because America "is looking more and more like Europe, with lower fertility, more measured levels of migration". Does that matter? For individual states, having fewer people undoubtedly hurts, because there are fewer taxpayers to pay for public services. But for the country as a whole Mr Chamie doubts that bigger is always better. He rejects the "pro-growth dogma" that having more consumers, workers and taxpayers makes sense, and notes the environmental costs of a bigger population. The US Census has set out scenarios for the population in 2060. Were America to return to being an exceptional place, open to high levels of immigration, its population could reach 447m. As a more normal rich country, less welcoming to immigrants, it could shrink to 320m.

Medically speaking: health, death and disease

Americans are driving less, so why are more of them dying in accidents?

In theory, one silver lining of the covid-19 pandemic should have been a decline in other types of death. People who spend more time at home drive less, and should therefore be less likely to die in car accidents. Data from the United States back up the first link in this chain: Americans drove 13% fewer miles in 2020 than in 2019. But less driving did not lead to fewer deaths. A statistical model from the National Highway Traffic Safety Administration (NHTSA) finds that deaths in the nine months to September 2020 increased by 5% year-on-year. Similarly, the National Safety Council, a non-profit, reckons that fatalities on roads rose by 8% in 2020. This implies that fatalities per mile driven rose by 24% – the biggest increase since 1924. Why are more people dying if fewer cars are on the road?

Although nationwide data lack detail, California publishes granular numbers on crashes and deaths. Drivers in America's most populous state have not become more accident-prone: collisions fell by 24%, more than miles driven (which fell by 13%). But the crashes that did occur were unusually deadly, causing at least one fatality 19% more often than in 2019. One factor was the reduced use of seat belts: even as collisions declined overall, the absolute number involving unfastened seat belts rose by 5%. Another factor was an increase in the severity of crashes in which reckless behaviour – such as alcohol or drug use, speeding or running red lights – was cited as a factor. Although the number of such collisions fell in line with the statewide average, the share that led to death rose by 23%, a greater increase than in other types of crashes.

The causes of a rise in risky driving are unclear, but responses to covid-19 probably contributed. Americans have been drinking more alcohol, and bingeing more often. Speeding is now more common: the number of tickets issued for going over 100 miles per hour (161km/h) doubled in California and Iowa, and the NHTSA reported that average speeds in a number of cities rose by 22%. Reductions in congestion and law enforcement may have emboldened drivers. In

California, rural areas, where roads are less crowded, account for a disproportionately high share of traffic deaths. But the gap in death rates between rural areas and cities shrank in 2020. That suggests that as urban roads emptied out, drivers started treating them more like rural ones, and stepped on the accelerator.

Why are rich countries so vulnerable to covid-19?

By the end of 2020, researchers had identified dozens of factors that can increase a person's chances of dying from covid-19, including hypertension, diabetes and obesity. But the biggest risk factor of all is being old. People in their 60s are twice as likely to die of covid-19 as are those in their 50s; the mortality rate of 70-somethings is higher still. Indeed, the probability of dying from the disease roughly doubles for every eight years of age. This helps explain why older, richer countries have fared worse than expected in the pandemic, compared with younger, poorer ones.

To estimate a country's vulnerability to covid-19, *The Economist* combined population data from the United Nations with age-specific infection fatality rates (IFRs) for the disease. The latter was estimated using data from Brazil, Denmark, England, Sweden, Italy, the Netherlands, Spain and parts of Switzerland and the United States. From these data we calculated an age-adjusted IFR: the probability that a randomly selected person from a given country would die if stricken with covid-19, assuming access to health care similar to that available in the sample countries.

We found that, all else being equal, Japan was more vulnerable to covid-19 than any other country in the world. With a median age of 48, it had an expected IFR of 1.3%. Italy, with a median age of 47, came next at, 1.1%. Europe as a whole had an expected IFR of 0.9%, by our reckoning, higher than that of America (0.7%), China (0.5%) or India (0.3%). Countries in Africa, meanwhile, had an age-adjusted IFR of less than 0.2%. Least vulnerable of all was Uganda, where the median age is just 17 and the expected IFR is a mere 0.1%, less than a tenth that of Japan or Italy.

Of course, all else is not equal. Other risk factors, such as obesity and smoking, vary significantly across countries. Health-care systems vary, too. Most important of all, fatality rates only matter to those who are infected in the first place, and infection rates vary from country to country, not least because of different

The curse of old age

Covid-19, estimated infection fatality rate
Adjusted for population age distribution*, %, 2020

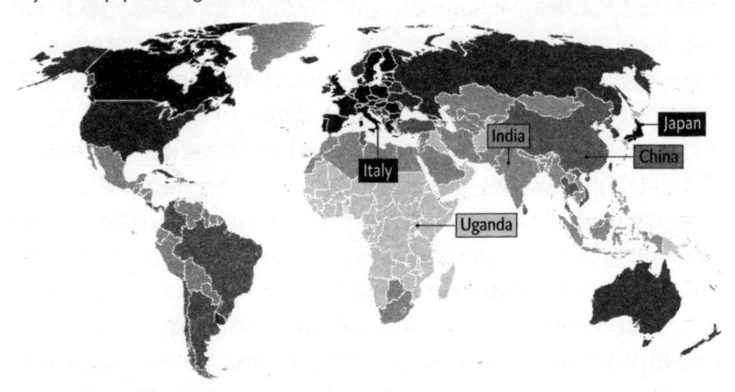

Sources: UN; "Covid-19 infection fatality ratio estimates from seroprevalence", N. Brazeau et al., MRC Centre for Global Infectious Disease Analysis, October 2020; *The Economist*

*Not adjusting for treatment options and underlying conditions

social-distancing rules and degrees of compliance with them. Nevertheless, knowing countries' age-adjusted IFRs is useful. It allows for fairer comparisons between countries. It also drives home the importance of keeping infections under control, especially in places with lots of older people, such as Japan and much of Europe.

How different kinds of vaccines work

Although lockdowns can keep the coronavirus at bay, vaccination provides the sustainable path out of the pandemic. More than 60 vaccines are either in development or current use against SARS-CoV-2, the virus that causes covid-19. All of those in use have the same ultimate result – of granting the body an enhanced ability to fend off viral attack – but the mechanisms they use differ considerably. So how do they work?

When the body becomes infected with a virus that it has never encountered before, the immune system begins a process of producing attack cells that can potentially destroy the intruder. This process takes both time and energy, because it involves considerable trial and error. It is one reason why people feel exhausted and ill for many days after contracting a new infection. If the body wins, the immune system remembers its successful strategy, so that future battles take much less time and symptoms are milder, or even non-existent. Vaccines are, in effect, military training programmes for the immune system. Rather than forcing the immune system to learn how to solve a problem during an actual attack, a vaccine sets up a mock attack for the body to practise on. Vaccines commonly use four types of mock attack, all of which are being deployed against SARS-CoV-2.

The oldest of these techniques is to present the immune system with the virus in a form that has been inactivated or significantly weakened, so that it cannot cause a full-blown infection. When immune cells detect such an intruder, they still engage in the arduous process of generating attack cells. But the immune system develops a memory of the event, which enables it to act quickly should it later suffer an attack from a similar virus. Both the Chinese Sinovac and Sinopharm vaccines use an inactivated virus. An alternative tactic involves injecting protein fragments that are structurally similar to SARS-CoV-2. These proteins are detected as foreign by the immune system, triggering an immune response. Novavax, an American developer, is using this technique in its covid-19 vaccine.

A more complex strategy involves inserting some of the DNA copied from SARS-CoV-2 into a version of a virus related to the common cold (known as an adenovirus) that can enter cells but has been neutered and cannot replicate effectively. Once inside the body, these engineered viruses feed the adenovirus DNA containing the copy of the SARS-CoV-2 DNA into cellular machinery that then leads the infected cells to produce lots of SARS-CoV-2 viral proteins and stick them on their surfaces, so that the immune system spots that something is awry. Crucially, the engineered virus also triggers an intruder alert system within these cells that draws immediate attention from the immune system. This stimulates the production of attack cells, creating an immune memory. This strategy is used by the AstraZeneca/Oxford, Johnson & Johnson and Russian Gamaleya vaccines.

The newest method involves researchers creating genetic instructions, in the form of RNA or DNA, that cause the body's own cells to generate harmless proteins that look like the distinctive proteins of SARS-CoV-2. As with the adenovirus approach, this method is potent because the RNA/DNA instructions turn human cells into viral protein production-centres that are all but certain to generate enough of the SARS-CoV-2 protein to trigger detection and hence to prime the immune system (which is not always the case with older vaccine technologies). The drawback with this technique is that DNA/RNA strands must be transported into the body in capsules that, though easily absorbed by cells, are exceptionally fragile and often require storage at very low temperatures. Both the Pfizer/BioNTech and Moderna/NIAID vaccines use this strategy.

In practice, all these approaches give the body a much better chance of mounting a successful defence against SARS-CoV-2 than it could otherwise. As for how well the different techniques can cope with the virus as it evolves, every one of them will need to be tweaked as new variants continue to emerge. The ease of tweaking, however, varies. Although RNA/DNA vaccines are the most cumbersome to transport and store, they are also the easiest to tinker with and may ultimately prove to be the most important if covid-19 evolves rapidly enough to evade the protection afforded by the first crop of vaccines.

Why the world desperately needs more nurses

When the World Health Organisation (WHO) declared 2020 the "Year of the Nurse and Midwife", to coincide with the 200th anniversary of Florence Nightingale's birth, few could have imagined that health-care workers would have such a turbulent time. They accounted for around 10% of covid-19 cases during 2020, according to the International Council of Nurses, a federation of national nursing associations. The disease has killed at least 1,500 nurses. But as well as highlighting their bravery and skill, the pandemic has exacerbated a problem with nurses: there aren't enough of them. England's National Health Service, for example, went into the pandemic with 40,000 nursing vacancies, a shortfall that is expected to reach 100,000 within a decade according to the Health Foundation, a think-tank. The WHO reckons that 6m extra nurses are needed globally.

To plug the gaps, Britain, like many rich countries, depends on immigration. Foreign-trained nurses account for 15% of all of Britain's nurses – over 100,000 of them, making Britain the largest destination country after America. But this workforce, once largely recruited from the European Union, is changing as employers turn to Asia. In 2019 fewer than 1,000 European nurses registered to work in Britain, compared with 9,389 four years earlier. The number of registrants from outside Europe, meanwhile, rose from 2,135 to 12,033 over the same period. Two factors caused this shift: Brexit, which slowed the arrival of European nurses to a trickle and prompted 5,000 to leave, and the loosening of immigration rules for nurses from outside the EU.

Britain's reliance on foreign nurses raises questions about the impact on poorer countries. The Philippines and India, the world's biggest exporters of nurses, and the origin of most of Britain's new foreign nurses, paint a complicated picture. Britain's government has a list of developing countries from which it does not recruit health workers, to prevent a brain-drain. The Philippines and India are on this list, but Britain has agreed on exceptions with their

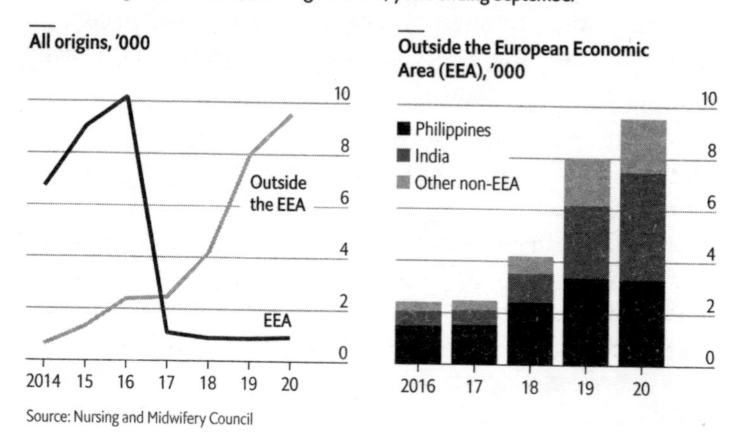

respective governments. Both countries are known for their private nursing schools which train aspiring migrants who are eager to go abroad. "Recruiting there is not regarded as unethical," says James Buchan of the Health Foundation. "There is an expectation that these nurses, mainly paying for their training, will move abroad." Indeed, many Filipinos and Indians rely on money that family members working abroad send home: remittances amounted to 9% of the Philippines' GDP last year.

But India and the Philippines face shortages, too, even though both countries train more nurses than they need. In Britain there are more than eight nurses per 1,000 people, compared with less than five in the Philippines and two in India. Nurses in these countries often face unemployment, low wages and poor career opportunities, all of which encourage graduates to move abroad. The WHO reckons the Philippines will face a shortfall of almost 250,000 nurses by 2030. International recruitment will continue as long as the world needs nurses, but filling vacancies will be difficult unless pay and working conditions improve, especially in poorer countries, according to the WHO. In July 2020 a report by

Britain's Royal College of Nursing found that 36% of its members were thinking about leaving the profession, an increase of 9% from 2019. After the threat of covid-19 has receded, rich and poor countries alike will be left wondering if they have enough nurses to weather another pandemic.

Why heatwaves are killing a record number of people

Until the summer of 2020 Verkhoyansk, a village in Siberia, was famous for being the coldest inhabited place north of the Arctic circle, with a record low of −67.8°C. In June 2020, however, it claimed another record: the hottest place north of the Arctic circle. The village recorded a high of 38°C, far above the usual summer peak of around 20°C. The freak temperatures were made 600 times more likely by man-made climate change, according to the World Weather Attribution project, a collaboration among climate researchers. As greenhouse-gas emissions in the atmosphere increase, extremes of heat such as that in Verkhoyansk are becoming more frequent. According to a report published in January 2021 in the *Lancet*, this is putting an increasing number of lives at risk.

A warming world poses many dangers, including flooding, pollution and the spread of disease, but the researchers identified heatwaves as one of the most striking. People already living in the hottest regions are the most exposed, but regardless of location, it is the elderly and those with pre-existing health problems such as heart disease who tend to suffer the most. The study combined population records with hourly temperature data from the European Centre for Medium-Range Weather Forecasts to track vulnerable people's exposure to heatwaves.

In 2019 people aged over 65 endured a combined extra 2.9bn days of extreme heat compared with a 1986-2005 baseline. This beats the previous record, set in 2016, by 160m days. India and China were among the worst affected, thanks to large populations and already hot regions. Heat-related deaths among the elderly, meanwhile, have risen by almost 54% between 2000 and 2018, according to the researchers' estimates. In 2018, the most recent year for which figures are available, heatwaves killed around 296,000 people over 65. Again, China and India suffered the most, with 62,000 and 31,000 deaths, respectively. Germany and America were the worst-affected Western countries, suffering around 20,000 deaths each.

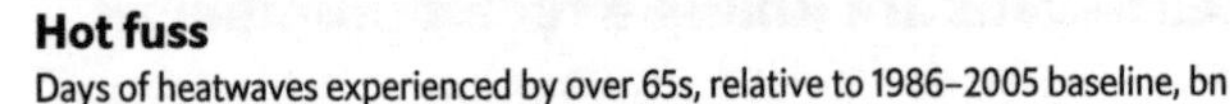

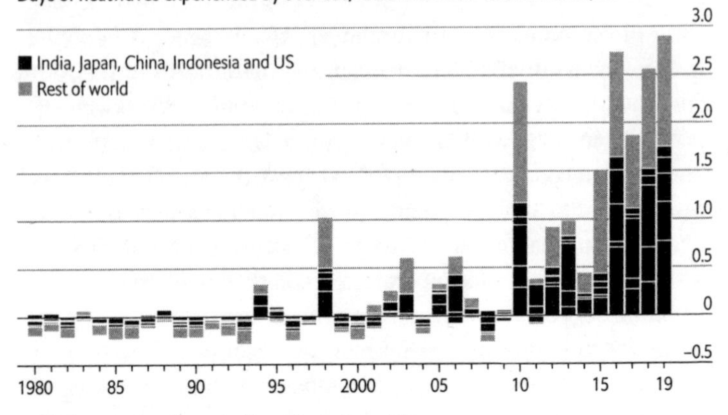

Source: The *Lancet* Countdown on health and climate change, 2020

A hotter planet has other costs too. In 2019 a total of 302bn working hours were lost to heatwaves, 103bn more than at the turn of the century, according to the study. India's productivity suffered the most, in part because of its large agriculture industry. In 2018 India had the second-highest agricultural output of any country, after China, but little farmland is irrigated, and much production depends on the monsoon. In high-income countries, such as America, construction was the industry worst affected.

The paper's authors note that the challenge that climate change poses to health-care systems is similar to that posed by covid-19, albeit less acute. As extreme-weather events become more frequent, countries need to identify the most vulnerable groups and invest in health care to ensure hospitals are not overwhelmed by climate shocks. But, as with the pandemic, prevention would be preferable to treatment. That will be hard. A separate report by the World Meteorological Organisation found that 2020 was one of the three warmest years on record, with average global temperatures of 1.2°C above pre-industrial levels. By 2024, that is expected to rise to 1.5°C.

Why is turkey meat becoming more popular?

Benjamin Franklin called the turkey a "Bird of Courage", saying that it was "a much more respectable Bird" than the eagle and a "true original Native of America". Franklin branded the patriotic eagle a "rank Coward" who steals his food from the fishing hawk and retreats from attacks by smaller birds. The turkey, by contrast, "would not hesitate to attack a Grenadier of the British Guards who should presume to invade his Farm Yard with a Red Coat on". (Despite being native to America, the bird got its name from the muddled belief in Europe that it originated from the Ottoman empire – ie, Turkey.) These days the doughty turkey has won the loyalty of millions of people for another reason – as a favourite festive-season meal. The average American gobbles 7.3kg of turkey meat per year, nearly twice as much as in 1970, and Europeans eat an average of 4.1kg. Why is it so popular?

Purchases of the bird are no longer confined largely to holiday seasons. In 1970 half of all turkeys consumed in America were eaten during festivities – Thanksgiving, Christmas and Easter. Today that share is down to 29%. Turkey consumption is also increasing elsewhere, notably in Asian countries that embrace Western-style diets. One reason for the growing appeal of turkey is the demand for low-calorie foods, says André Laperrière of Global Open Data for Agriculture and Nutrition, a non-profit group. A 100-gram portion of turkey contains, on average, fewer calories (189) than the same amount of beef (250), pork (242) or even chicken (239), whose growing popularity is the result of similar factors. Not only is turkey meat healthier than rival products, it is also cheaper. The retail price of frozen turkeys in America in 2019, according to the US Department of Agriculture, averaged $1.52 per pound (454 grams), against $5.44 per pound for fresh beef for roasting.

Turkey production is also less damaging to the environment than other meats (though plant-based food is even less damaging). For every pound of beef, six pounds of feed are required, whereas a pound of turkey requires just two pounds of feed. And while producing a kilo of beef requires 15,000 litres of water (not just to keep the animals' thirst at bay, but to produce the feed that fattens them), a kilo of poultry (turkeys included) requires 4,300 litres.

Like many industries, the turkey-meat trade was not spared by the covid-19 pandemic. Restrictions on social gatherings meant that fewer families got together for Thanksgiving and Christmas in 2020, denting the demand for bigger birds. "It might take a couple of years before the market bounces back to normal," says Mr Laperrière. But, he says, "I'm confident it will flourish over the next decade." The Bird of Courage will, no doubt, rule the roost again.

How the Japanese boosted their longevity by balancing their diets

Tanaka Kane is one of humanity's great outliers. On January 2nd 2021 she became the third person ever to turn 118, according to the Gerontology Research Group, a global team of academics. She was also the first citizen of Japan to reach 118 – but is unlikely to be the last. The country has the world's longest life expectancy, and more than 80,000 centenarians. Mrs Tanaka is an outlier for another reason, too. She claims to love chocolate and fizzy drinks, setting her apart from most of her compatriots. Japan has long had one of the lowest sugar-consumption rates in the OECD, a club of mostly wealthy countries. The unusual longevity of the Japanese is often credited to diet. Yet the idea that the country has extended lifespans by entirely avoiding the West's sinful culinary delights may be too simple. In fact, studies suggest that one key to its success may be that its people's diets have shifted over time towards Western eating patterns.

Japan was not always a longevity champion. In 1970 its age-adjusted mortality rates were average for the OECD. Although its levels of cancer and heart disease were relatively low, it also had the OECD's highest frequency of cerebrovascular deaths, caused by blood failing to reach the brain. Between 1970 and 1990, however, Japan's cerebrovascular mortality rate fell towards the OECD average. This, combined with world-beating numbers on heart disease and fewer strokes, meant that Japan soared up the longevity league table. How did Japan overcome its cerebrovascular woes? Some of its gains simply mirror better treatments and reductions in blood pressure around the world, notes Thomas Truelsen of the University of Copenhagen. But another cause may be diet.

Japan largely banned meat for 1,200 years, and still consumes relatively little meat and dairy. Too much of either can be damaging, because they contain saturated fatty acids, which correlate to heart disease. Studies have also tied eating lots of processed red meat to a greater risk of stroke. But too little may be unwise as well, because

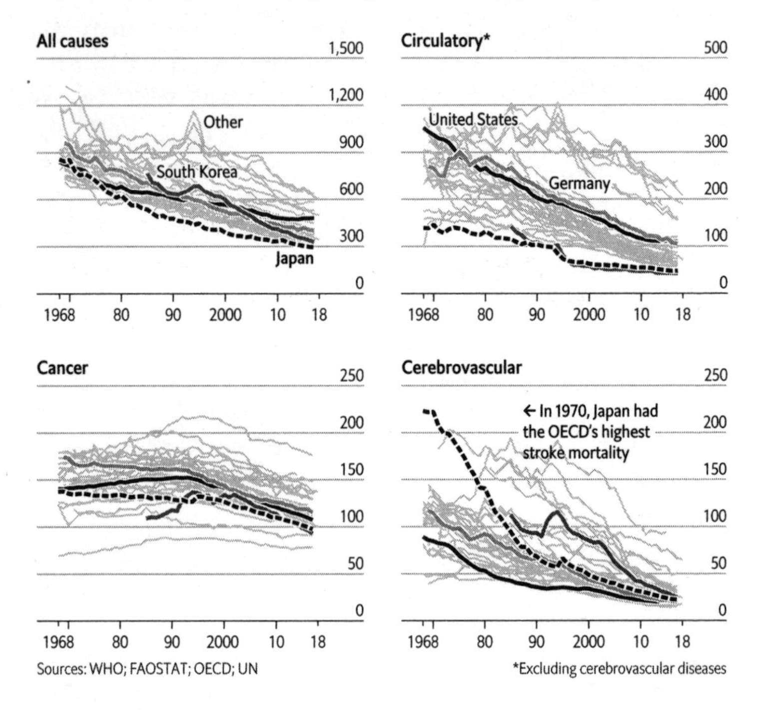

meat and dairy products provide cholesterol that may be needed for blood-vessel walls. In a study of 48,000 Britons, vegetarians were unusually resistant to heart disease, but prone to strokes.

In theory, a dearth of animal-based food could have contributed to Japan's historical cerebrovascular mortality. From 1960 to 2013, as the country's deaths from strokes tumbled, its annual meat intake rose from near zero to 52kg per person (45% of America's level). Tsugane Shoichiro of the National Cancer Centre in Tokyo says that his compatriots may need meat and dairy to keep their blood vessels robust – though not so much that those vessels get clogged. Some empirical evidence supports this view. One paper from the 1990s found that the parts of Japan where diets had changed the most also had the biggest drops in cerebrovascular mortality. Another study, which tracked 80,000 Japanese people between 1995 and 2009, showed that strokes were most common among those who ate the least chops and cream. Although Japan's decline in cerebrovascular deaths could stem entirely from other causes, these data suggest that nutritional shifts may have helped.

The unhappy irony is that Japan's health gains, paired with a low birth rate, threaten its economy. By 2060, 40% of Japanese could be 60 or older. That would yield more birthday cakes with 118 candles – but fewer great-grandchildren to help blow them out.

Why playing video games in lockdown is good for your mental health

When the pandemic put a stop to normal life, many people switched on their games consoles. Video-game internet traffic in America rose by 75% within a week in March 2020, as many states went into lockdown, according to Verizon, a telecoms giant. Sony struggled to ship enough of its latest PlayStation consoles to meet demand. And one of the year's most popular games, "Animal Crossing: New Horizons", sold more than 31m copies between March and December 2020 (it is close to becoming the biggest-selling game for Nintendo's latest console). Players of the game boasted on social media of spending hundreds of hours on an island full of talking creatures. Too much screen time? On the contrary. A study by researchers at the University of Oxford reckons that this was time well spent.

The researchers collected data on the gaming habits and mental health of players of two games, in Britain and North America. Some 2,537 of the participants, with an average age of 31, played "Animal Crossing: New Horizons" and 468, with an average age of 35, played "Plants vs Zombies: Battle for Neighborville". The researchers worked with the developers of the games to record when participants played during August and September (when asked to self-report, people tended to overestimate their playing time). They also measured the players' well-being, by surveying how often they reported experiencing each of six positive and six negative feelings over the previous two weeks.

The researchers found that people who played the games for longer reported feeling better, on average, than those who barely played at all. They stopped short of claiming that playing time directly affected well-being, noting that people who already felt good might have been more inclined to play. Nor did they look at the effect of playing on people who are not habitual gamers. They did find, however, that certain feelings provided by video games, such as a sense of freedom and competence, improved the players'

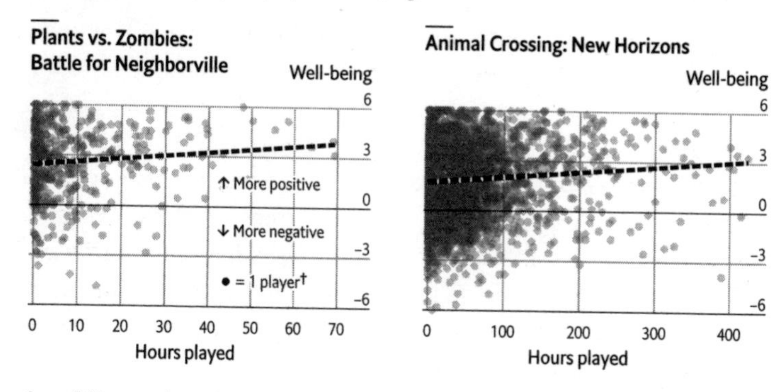

Source: "Video game play is positively correlated with well-being", N. Johannes, M. Vuorre and A. K. Przybylski, PsyArXiv, 2020

*Net well-being score after two weeks. 6=most positive, −6=most negative
†Adult players in the United States, Canada and Britain

sense of well-being while they played. A greater feeling of social connection from playing with others in the game, crucial when friends could meet in person, also boosted their mood.

Many policymakers do not share this positive view of video-gaming. In 2020 Britain's government announced plans for a regulator to protect children from "excessive screen time", and the World Health Organisation has identified "gaming disorder" as addictive behaviour. There is a lack of robust evidence for many of the supposed harmful effects of video games, according to Andrew Przybylski, one of the academics behind the study. More research is needed to ensure that gamers, especially young ones, are adequately protected without spoiling their fun (the gamers in the study were well into adulthood). But worries about too much gaming, it seems, may be overblown. There are many worse ways to pass the time during lockdowns.

Follow the money: economical, with the truth

Why China's economic centre of gravity is moving south

"Don't invest beyond Shanhaiguan" has been a popular quip in China for years, referring to a pass in the Great Wall that leads to the north-eastern rust belt. Online pundits have recently updated the maxim to "Don't invest outside the Southern Song", a dynasty that fell almost 750 years ago whose territory was roughly the same as China's southern half today. The joke contains a kernel of truth: China's southern provinces are outperforming northern ones in nearly every economic dimension. In January 2021 China announced that its GDP had grown by 2.3% in 2020, making it one of the few countries to expand in the virus-blighted year. The recovery was unbalanced, with factories at full throttle but consumption subdued. Things should improve after the pandemic ends, as people move around more. The north–south imbalance, by contrast, was worsening before covid-19 – and is likely to outlast it.

The south's share of GDP has risen to 65%, the highest on record, and up from 60% five years ago. Some of that is a matter of luck. The north, home to China's largest coal mines and oil reserves, was caught out by the fall in commodity prices after 2013. It also boasts big industrial firms, from steelmakers to chemical giants. China's shift away from construction-fuelled growth towards consumption and services has hurt.

Other factors are at work, too. Partly because of their industrial heritage, northern provinces have strong command-economy tendencies. An analyst with a credit-rating agency asked officials in Tianjin, a struggling northern city, whether they would allow more defaults by state-owned firms. "Market solutions work best in the south," was the reply, he says. Officials in the north tried harder to goose up growth, to the region's detriment. In 2013, the peak of China's building frenzy, investment in assets such as roads and factories accounted for an eye-watering 66% of GDP in the north versus 51% in the south. Local governments in the south have been more hands-off. China's two most dynamic regions, home to its

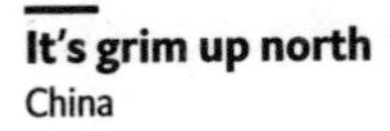

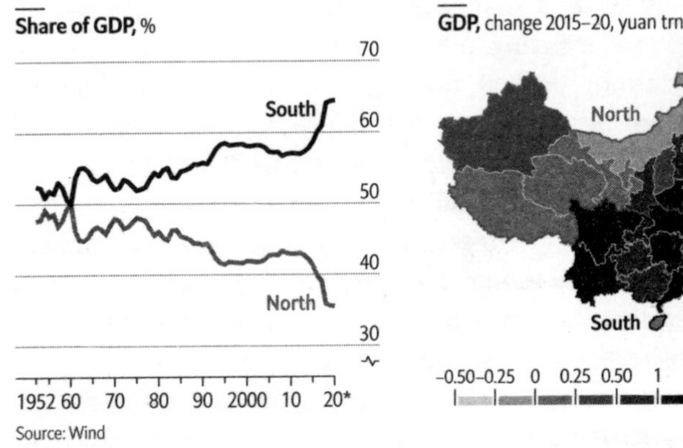

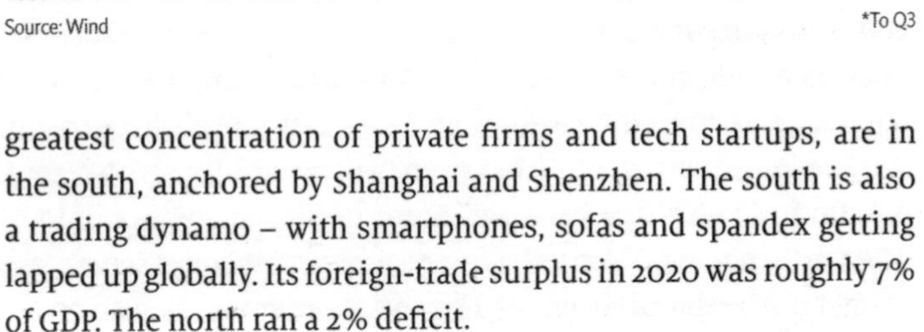

greatest concentration of private firms and tech startups, are in the south, anchored by Shanghai and Shenzhen. The south is also a trading dynamo – with smartphones, sofas and spandex getting lapped up globally. Its foreign-trade surplus in 2020 was roughly 7% of GDP. The north ran a 2% deficit.

Adding insult to injury, the north was also disrupted more by sporadic covid-19 outbreaks in late 2020. Millions of people in Hebei, a province that makes about a quarter of China's steel, were locked down. Geography was part of the problem: a harsher northern winter makes the virus more transmissible. But even without the pandemic, the north is increasingly out in the cold.

Why Black Friday isn't what it used to be

In 2020, Oxford-dictionary lexicologists declared "blursday" one of the words of the year. Pandemic-induced date confusion was not limited to self-isolating households, however. Best Buy, a large American electronics retailer, declared that "Black Friday isn't just one day this year – it's months long." Black Friday, the start of the pre-Christmas shopping season, has long been a bonanza for American retailers. The term itself is often credited to Philadelphia's policemen, who used it to describe the pandemonium caused by suburban shoppers and tourists thronging the city ahead of the annual Army–Navy American-football game on the Saturday after Thanksgiving. By the 1980s shops recognised the branding opportunity – and began marking the occasion with deep discounts and "doorbuster" deals to pull people from their turkey-laden tables into shopping aisles.

These days retailers make one-fifth of their holiday revenue, defined as sales in November and December, in the five days between Thanksgiving to Cyber Monday, a name invented in 2005 by the National Retail Federation (NRF), a trade group, in recognition of an online-sales bump on the first working day after the holiday weekend. (This bump was attributed to the theory that many shoppers had faster internet connections at work than at home, and therefore found it more convenient to shop from their desks.) In an ordinary year, Black Friday typically generates more than twice the foot traffic of other annual shopping sprees in America.

In 2020, though, covid-19 made many shoppers reluctant to elbow their way to cut-price wedding gowns or giant televisions. This prompted many large retailers, including Best Buy, to stretch Black Friday from a frenzied 24 hours to several weeks. Walmart, Target and other big-box retailers announced discounts on holiday items as early as October 11th. Instead of mall Santas and mistletoe, they offered refurbished online interfaces, generous return policies and expanded options for kerbside pickup. The pandemic-induced collapse of travel and other "experiential" spending meant

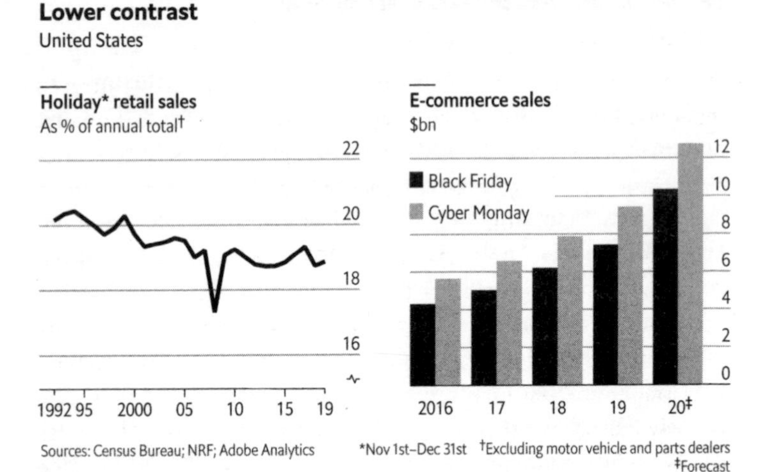

customers had money to spend on other things, noted Jill Standish of Accenture, a global consultancy. Holiday sales grew by 8.3% compared with 2019, as friends and families used gifts to show long-distance appreciation.

Even so, the broader picture is one of long-term stagnation. The holiday season's share of annual retail sales has been around 19% throughout most of the past decade, down from more than 20% in 1992. Online shopping offers perpetually low prices, making one-off discounts somewhat less exciting. Even without a pandemic, Black Friday's in-store stampedes no longer look that appealing. The day's internet sales have been rising (though Cyber Monday has digitally outshone it since at least 2016). In 2019 a third of the Black Friday's $23bn trade happened online. Now the online share could be closer to a half. The idea of squeezing all your bargain-hunting into one day is falling out of fashion. Since 2019 Amazon Prime Day, the e-commerce giant's signature shopping event, has lasted 48 hours. And in 2020 Singles' Day, a Chinese extravaganza which normally falls on November 11th, lasted a full 11 days.

What is the African Continental Free Trade Area?

Kwame Nkrumah, Ghana's first president, led his country to independence in 1957, an achievement that helped inspire many African states to break ties with their colonial powers. Six years later he spoke at the inaugural conference of the Organisation of African Unity. "The people of Africa are crying for unity," he declared. Nationalist though he was, Nkrumah still dreamed of a United States of Africa. He envisaged harmonised systems, dismantled boundaries and more intra-African trade. Some six decades later, his dream came a little closer to reality when the African Continental Free Trade Area (AfCFTA) came into effect at the start of 2021. The framework ties together the biggest number of member countries of any trade agreement since the establishment of the World Trade Organisation in 1995.

The agreement was signed in 2018, but the idea of building a single market for goods and services in Africa had been proposed by the African Union (AU) six years earlier. It is one of the AU's flagship projects and part of its plan to turn the continent into a global economic power over the next few decades. All African countries but one have signed the agreement – Eritrea shunned it in favour of existing regional economic deals – and 36 have ratified it. Some economists predict that the AfCFTA is a game-changer. The World Bank estimates that it could boost Africa's GDP by 7% – almost $450bn – by 2035, in part by reducing import tariffs, but more importantly by eliminating non-tariff barriers. It should also deliver notable benefits in how income is distributed, potentially lifting some 30m people out of extreme poverty and 68m out of moderate poverty.

Such a deal was clearly needed if African businesses were to achieve greater economic integration. Africa is still struggling to detach itself from its colonial past. It remains heavily reliant on external trading partners, notably in Asia and Europe. A paltry 16% of Africa's trade is intra-regional, compared with 60% for Asia and

68% for Europe. The continent already has several regional trade agreements, but high tariffs divide the different pacts. Exporters from the six countries of the East African Community, for instance, face an average tariff of 16% on products they send to the northern Arab Maghreb Union. Regional trade blocs continue to exist under the continental free-trade agreement. But it provides an Africa-wide regulatory framework to harmonise intra-continental trade.

The AfCFTA is intended to tackle such obstacles to trade, starting with the removal of tariffs on 90% of goods within five to ten years. Non-tariff barriers, such as complex customs procedures, excessive bureaucracy and corruption, also pose considerable challenges. Take the example of east Africa. Mozambican customs officers charge drivers from eSwatini 500 meticais (around $7) for vehicle inspections that are in fact rarely performed. To help overcome non-trade barriers, an online portal has been set up as part of the AfCFTA, through which African businesses can report any such problems they face. Complaints are handled by nominated government officials, who are meant to take action to remove the barriers.

Yet although the pact is in place, not all countries are in a position to benefit from it. Some are hampered by poor transport networks and inefficient customs procedures. Others are scarred by years of violence and regional disputes. The AfCFTA's success depends on political co-operation. African leaders know that change will not happen overnight. Wamkele Mene, the secretary-general of the AfCFTA, said the task was "daunting", but also crucial to ensure Africa's economic stability and sustainability: "We're not going to get another opportunity to integrate our market."

Why money buys happiness, but euphoria comes dear

Money might not guarantee happiness, but income does tend to correlate with contentment. In a Gallup poll carried out in 2020, residents in the top 10% of countries by GDP per person scored their life situation as seven out of ten on average, compared with just four out of ten for those in the bottom 10%. But what difference do individual earnings make? A paper published in January 2021 by Matthew Killingsworth of the University of Pennsylvania found that happiness continues to increase even as income ascends to plutocratic proportions, with two caveats. First, the more happiness you want, the more expensive it gets. And second, money is not nearly as important as other factors.

In 2010 Daniel Kahneman and Angus Deaton, both of Princeton University, found that happiness, as measured by people's own perception of their emotional well-being, levelled off when annual incomes reached around $75,000 (or $90,000 today). Mr Killingsworth's subjects, by contrast, mostly got happier as they earned more. The catch is that the next dollar a person makes will cheer them slightly less than the last one did. The average difference in life satisfaction between two people earning $40,000 and $80,000 is about the same as that between two earning $80,000 and $160,000. And Mr Killingsworth found that only a small percentage of the overall variation in happiness was explained by differences in income. Previous research has found that health, religion, employment and family are all important. In other words – for those who need to be told – there's more to life than money.

Part of the explanation for the differences between the new research and the old lies with methodology. Earlier research asked participants to think back on their days and remember how they felt. In contrast, Mr Killingsworth's subjects received alerts on their smartphones asking them to rate their current mood and their life satisfaction. Mr Killingsworth also used a more granular scale and surveyed each subject more often than past research,

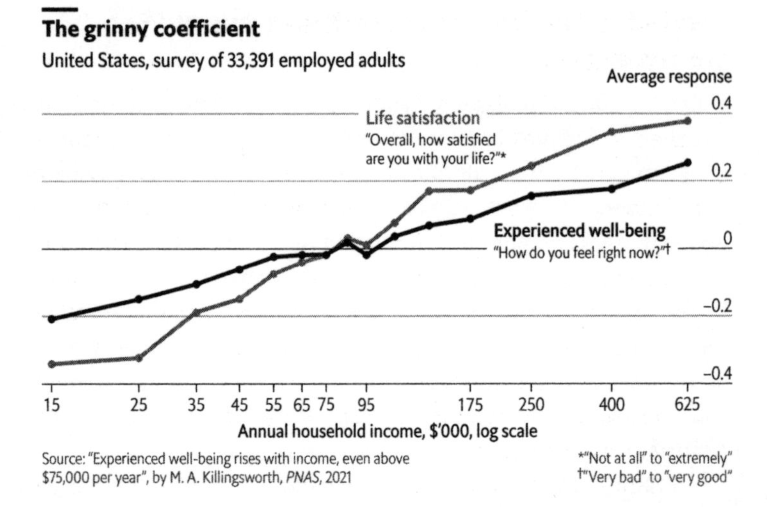

potentially making for more precise data. There is also a matter of interpretation: when visualised differently using an unlogged scale, the diminishing returns to increasing income found by Mr Killingsworth look not dissimilar to the levelling-off reported before.

Such research offers two lessons for politicians who want to improve people's moods. First, helping the poorest is a bargain. In happiness terms, a dollar goes further for someone earning $20,000 a year than for someone on $40,000. Second, economic growth, much maligned as a yardstick of progress, is important as long as it doesn't come at the expense of other measures of well-being. Happiness depends on many factors, but a more prosperous future is probably a more contented one too.

What is the fuss over central-bank digital currencies?

Central-bank digital currencies are coming. China has started large-scale trials of the digital yuan. European officials want to launch a digital euro by 2025. America's Federal Reserve is studying whether to follow suit. The Bahamas has already put its version, the "sand dollar", into circulation. This sudden rush arouses both excitement and confusion. A central-bank digital currency, or CBDC, sounds like a newfangled monetary innovation, or perhaps an official take on bitcoin. But it is not obvious what the point is. Many people already use digital currency, whether in mobile apps for payments or on bank websites for transfers. What are these new digital currencies, and why are central banks creating them?

CBDCs are simply a digital version of cash – the physical money issued by central banks. In most countries, their design will probably resemble existing online payment platforms, with one key difference: money held on a CBDC app or website will be equivalent to a deposit at the central bank. The main motivation for central banks is to limit the risks inherent in the global shift to cashless payments. They are responsible for the safety of the monetary system, the most basic element of which is guaranteeing that people can use cash to buy things. But in a world dominated by Apple Pay or Alipay, everyday transactions would depend on private companies rather than on central banks. Cryptocurrencies like bitcoin and "stablecoins" (which peg their value to the dollar or other assets) like Facebook's Diem (formerly Libra) are also a threat, potentially chipping away at state authority. CBDCs would give central banks a stronger presence in online payments. That would reduce the risk to financial stability posed by reliance on purely private payment systems. And some people might like having the option to keep at least some of their digital cash on a trusted, official platform.

CBDCs are not just a means of defending central banks' turf, however. Central banks also see opportunities. Cashless transactions make for faster, more reliable payments and are less susceptible to counterfeiting. Issuing digital cash is cheaper than minting coins, so long as it is protected against hacking. Officials would also have an easier time monitoring how digital money is used, making it harder to fund criminal activities. In poorer countries central banks hope that digital currencies will bring unbanked citizens into the financial system, boosting economic development. Central banks that issued CBDCs could also gain new powers. One impediment to negative interest rates at the moment is that savers can switch to hard cash, which has a de facto interest rate of zero. In a cashless world central banks could, in theory, program a digital currency to have negative rates. They could also use mobile wallets to make direct cash injections more efficiently – for example, to send money quickly to residents of a region struck by an earthquake.

These ambitious uses of digital currencies are still far off in the future. For years to come, central banks will continue to provide banknotes alongside e-wallets, recognising that many people still want to hang onto hard cash or are unwilling or unable to use smartphones. Central bankers are also cautious by profession. Jerome Powell, the chairman of the Federal Reserve, has said that America would prefer to "get it right than to be first" with a digital currency.

In fact, central bankers already have plenty of experience in managing what are known as "wholesale" digital currencies: the reserve deposits of commercial banks at central banks. The real question is how to design "retail" digital currencies for the public. One option is for central banks to offer their own official mobile apps. Another is for people to use their preferred payment apps, as they do now, but to have some cash on them directly tied to an account at the central bank, rather than sitting on the payment provider's balance-sheet. Some policymakers worry about "digital bank runs", which could occur if people convert their savings into CBDCs in a panic, draining banks of liquidity. Caps on CBDC holdings or transfers might be a useful safeguard, at least to begin with. Sorting out all these details will take time. But the general direction is clear. As cash disappears, official digital currencies will emerge, and the links between people and central banks are likely to become stronger.

Why people from poor countries pay more for visas

As borders start to reopen after more than a year of stop–start coronavirus restrictions, some travellers are beginning to plan holidays and work trips again. In addition to packing sunscreen or getting their PowerPoint presentations ready, many will have to sort out their visas. For some unlucky travellers, this will prove costly. A working paper published in June 2020 shows that visa prices vary enormously depending on the traveller's passport. Researchers from the European University Institute collated prices for 85,000 different visas to and from virtually every country in the world – including those intended for tourists, business travellers, foreign workers and students. Taking each country in their data set in turn, they noted how much it would cost a citizen of that country to buy a visa to numerous other countries and worked out an average. They found that people from richer countries had to pay considerably less for a visa than those from poorer ones.

North Africans pay more than twice as much for a tourist visa than North Americans. And when the cost of an average tourist visa was compared with average national income, the disparity between rich and poor countries became even starker. Whereas North Americans have to work for less than half a day, on average, to pay for a tourist visa, sub-Saharan Africans have to work for 19 days. Such differences are sometimes the result of costly levies. Ecuador charges $450 for would-be travellers from some poor or conflict-ridden countries, meaning Ethiopians must stump up seven months' national-average salary for the privilege of visiting. Waivers allowing citizens to travel visa-free are also granted to rich countries more often than to poor ones.

Is racial discrimination to blame for the sky-high fees charged to African tourists? Probably not. The researchers tested this by adding two variables to their model that, according to previous academic research, are important for explaining visa waivers – economic prosperity, as measured by income per person, and

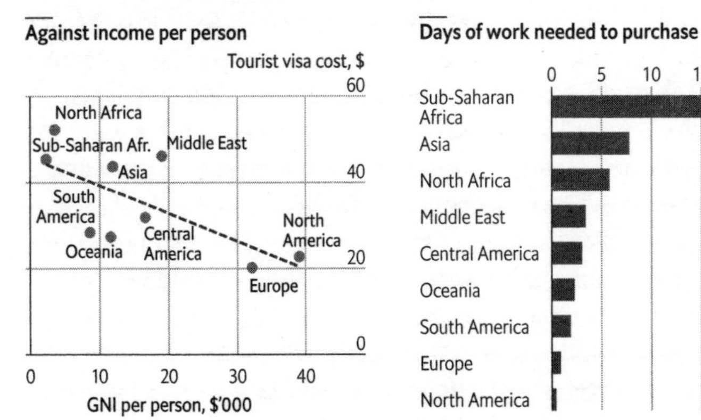

quality of democracy, as measured by the Economist Intelligence Unit, our sister company. They found that when these variables were included, the additional costs associated with travelling from Africa disappeared. In fact, controlling for these factors, travelling for work from Africa looked relatively cheap compared with costs for travellers from poor countries on other continents.

Those hoping to escape these visa costs with the help of the private sector may be disappointed. The researchers compared the prices of 3,000 visas from a leading online agency, touristvisaonline.com, with those found on official government websites. They found that the agency was, on average, 90% more expensive (excluding fees). The amounts by which the firm overcharged its clients, relative to governments, tended to be round numbers, such as 50%, suggesting it was deliberate. As seasoned travellers will know, the kindly soul who offers to help you navigate local bureaucracy very often has an angle.

When bribery pays – and when it doesn't

Corporate bribery is hardly new. Surveys show that between a third and a half of companies typically claim to have lost business to rivals who won contracts by paying kickbacks. But such perceptions-based research has obvious limitations. A study published in December 2020 in *Management Science* takes a more rigorous approach, and draws some striking conclusions. Raghavendra Rau of Judge Business School at the University of Cambridge, Yan-Leung Cheung of the Education University of Hong Kong and Aris Stouraitis of Hong Kong Baptist University examined nearly 200 prominent bribery cases in 60 countries between 1975 and 2015. For the firms doing the bribing, they found, the short-term gains were juicy: every dollar of bribe translated into a $6-9 increase in excess returns, relative to the overall stockmarket.

That, however, does not take account of the chances of getting caught. These have risen as enforcement of America's 43-year-old anti-bribery law, the Foreign Corrupt Practices Act (FCPA), has been stepped up and other countries have passed similar laws. The number of FCPA cases has risen sharply since the financial crisis of 2007-09, according to Stanford Law School. It dipped a bit under President Donald Trump, who criticised the FCPA for hobbling American firms overseas, but remained well above historic levels. Total fines for FCPA violations were $14bn in the four years to the end of 2019, 48 times the amount in the four years to the end of 2007.

The authors also used their data to test 11 hypotheses that emerged from past studies of bribery. They found support for some: for instance, that firms pay larger bribes when they expect to receive larger benefits, and that the net benefits of bribing are smaller in places with more public disclosure of politicians' sources of income. But the researchers punctured other bits of received wisdom. Most strikingly, they found no link between democracy and lower levels of corruption. This challenges the "Tullock paradox", which holds that firms can get away with smaller bribes in democracies because

Brown envelopes, big cheques

United States, Foreign Corrupt Practices Act

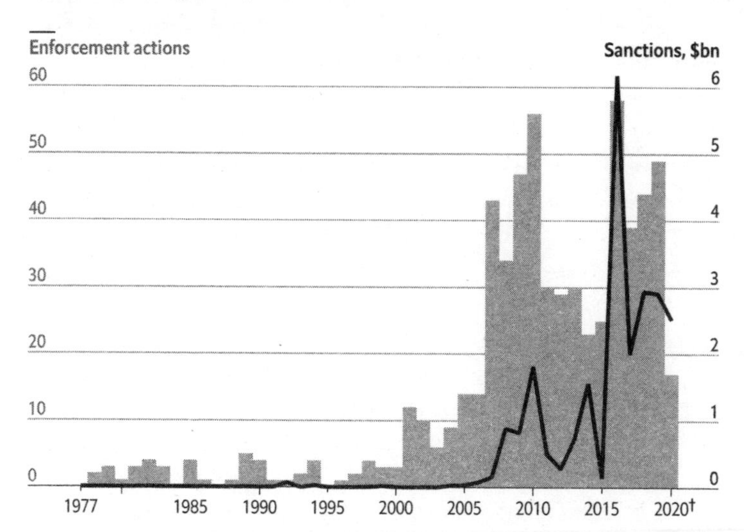

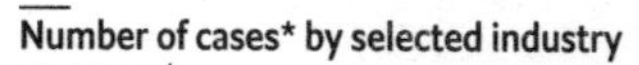

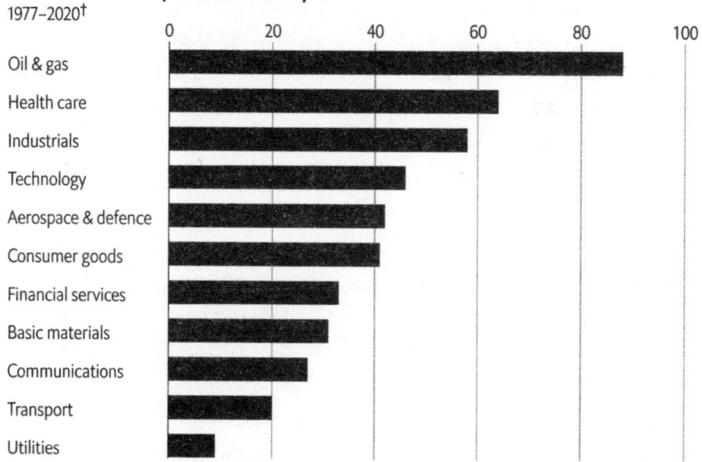

Sources: Stanford Law School; Sullivan & Cromwell

*Investigations and enforcement actions †To August

politicians and officials have less of a lock on the system than those in autocratic countries, and so cannot extract as much rent. These findings will surely be of interest to corruption investigators – and, no doubt, to unscrupulous executives too.

Speaking my language: words and wisdom

Why Japanese names flipped (for Western readers) in 2020

On January 1st 2020 a minor lexical revolution rolled through Japan. A new decree ordained that official documents should reverse the order of Japanese people's names when they are written in the Latin alphabet. Previously in, say, English documents, Japanese names had been written with the given name first, using the Western practice. But under the new rules the family name now comes first and, to banish any ambiguity, may be entirely capitalised. One backer of the change was the prime minister at the time – who, as a result of the change, is now referred to by *The Economist* as Abe Shinzo, rather than Shinzo Abe.

Like other newspapers, *The Economist* had long followed the convention of writing Japanese names in the Western order (while scholarly publications tended to use the Japanese order). But, as is common in East Asian cultures, in Japanese the family name always comes first. Japanese conservatives do not see why they should say their names backwards just for the convenience of Western minds too lazy to grasp a basic facet of Japanese culture. Some 59% of Japanese in an opinion poll favoured the switch to surname-first.

Yet there is an irony. Japan first decided to put given names first when talking to foreigners way back in the 1870s, during the Meiji era. It was actually a gesture by nationalist reformers who wanted to keep Western imperialists at bay. Japan, they argued, could keep its independence only by abandoning the paternalism of Confucius, imported from China, which had long governed society and family life. Instead it should rapidly learn modern Western ways in everything from military affairs to education, both to ward off Western powers and to impress them. Adopting English name order was a tiny part of the package. Reformists had drunk deeply of the social Darwinism then prevailing in the West, which taught that only the strongest societies would survive. One, Mori Arinori, even proposed adopting English as Japan's language.

In 1885 a friend of Mori's, Fukuzawa Yukichi, penned a polemic,

"Goodbye Asia", arguing that Western civilisation was like measles: if it didn't kill you, it would make you stronger and should be embraced. He said the static cultures of China and Korea would make those countries more vulnerable to Western conquest. He urged Japan to cut its spiritual and civilisational ties with them. It was, with hindsight, a small step from there to a sense of Japanese exceptionalism, and then to militarism. Chinese reformers and revolutionaries were later inspired by Fukuzawa and his like to reform or abandon the Chinese language. The great novelist Lu Xun argued in the 1920s that the Confucian ideologies holding China back were being subconsciously reinforced by the archaic Chinese used in writing. He called for a new vernacular. Chen Duxiu, a co-founder of the Communist Party, urged the use of Roman letters to replace the thousands of Chinese characters.

As for Japan, its defeat in the second world war aggravated what a Japanese writer, Mizumura Minae, in *The Fall of Language in the Age of English*, deems an inferiority complex among Japanese intellectuals regarding their own tongue. One novelist, Shiga Naoya, even blamed the Japanese language for the start of the war (he advocated a switch to French). But in the 21st century, with China on the rise and America proving a wobbly ally, Mr Abe and his allies saw the change in the naming rules as one way for Japan to stand tall again and celebrate tradition.

Still, asks Jeffrey Kingston of Temple University in Tokyo, is name order really "the big naming question facing Japan in the 21st century"? Surely, he says, a bigger issue is the official ban, dating from 1896, on married couples retaining separate family names – a rule that in practice means wives nearly always take their husbands' names. After a failed legal challenge in 2015, that rule seems likely to remain for the foreseeable future. In February 2021 Marukawa Tamayo, Japan's minister for women's empowerment and gender equality, said she was opposed to any change.

Where did the word "robot" come from?

Who invented Siri or Alexa? The obvious answer is the software wizards at Apple or Amazon who devised the voice-activated digital assistants that serve as virtual butlers in people's homes. But who originally hatched the idea of ever-ready non-human helpers that could "speak, write and do arithmetic" – hi-tech entities endowed with "amazing memories" and total recall but without the ability to "think of anything new"? Those quotations come from Karel Capek's play *R.U.R.* ("Rossum's Universal Robots"), which had its premiere in Prague on January 25th 1921. Stories about man-made beings, animated statues or high-tech automata have long beguiled humans, from ancient times to Mary Shelley's *Frankenstein*. However, Capek created one of the first fully imagined artificial servants equipped not only with brawn but a facsimile of a brain. His dystopian satire of the factory-assembled slaves who revolt against their human overlords introduced the word "robot" – derived from an old Czech word for forced labour performed by serfs – to a world that soon adopted it in art and technology alike.

Within two years, *R.U.R.* had reached the stage in London and New York. Since then, it has driven a century of debate about automation, alienation and the promise (or threat) of new technologies that mimic human capabilities. Capek never conceived of his "robots" as clanking tin contraptions but as bio-engineered creatures. They begin with no "will" or "soul", but later acquire something like sentience, empowered by a movement "to set the robots free".

A century on, these original Czech robots look less like ancestors of the cute or scary mechanisms in science-fiction plots; rather, they seem like avatars of modern anxieties about artificial intelligence and its potential. For a start, they exist not as pre-set hardware but quasi-organisms spun in an island factory from living "slime", or "some sort of colloidal jelly". Domin, the robot plant's boss, mistakes their nature. He sees only the low-cost, high-output assembly-line kit ("the cheapest workforce you can get"). For him, as for many industrial utopians, robotics means liberation from drudgery and want as "everything will be done by living machines".

The robots, though, will have other ideas – once they have ideas at all. The crucial first step towards self-awareness comes when Dr Gall, a research director, reprograms them with a sense of pain, purely as "an automatic protection against injuries". Spurred on by the solidarity of Helena, a sympathetic visitor, the robots soon exhibit "something like rage or defiance". They aspire to dignity and freedom, but catastrophically conclude that "to be like people, it is necessary to kill and to dominate". Cue the final showdown as the leader of the robots, Radius, proclaims that "the era of man has come to its end." A coda, though, hints that the robot "soul" may evolve in all-too-human ways.

Capek was an anti-totalitarian liberal allied to the first president of independent Czechoslovakia, Tomas Masaryk. His vision of the regimented masses who revolt, but mimic the violence of their oppressors, clearly belongs to the aftermath of the Russian revolution. A twist in which the embattled humans re-engineer "universal" robots as localised models, to make them "strangers to each other", shows his dismay at rising nationalism. The search for an escape from "the slavery of labour" in *R.U.R.* stems as well from the working world of Czechoslovakia after 1918. The new state inherited 80% of the factories of the old Austro-Hungarian empire; it ranked among the world's ten most industrialised economies. The stress, fatigue and injury that Domin wants robots to eliminate meant a heavy workload for the Workers' Accident Insurance Institute in Prague. One of its most respected senior officials dabbled a bit in writing himself. His name was Franz Kafka.

Both Kafka, who dealt every week with the mutilated casualties of industry, and Capek understood why human ingenuity would seek release from the "labour and anguish" of workshop and factory. But *R.U.R.* leapfrogs the era of automated liberation to preview two challenges that technology still confronts today. First, how much autonomy do humans wish our smart machines to have? And second, what should mankind do instead of the increasingly sophisticated tasks undertaken by robot aids? Alquist, Capek's down-to-earth builder in *R.U.R.*, advocates a return to honest toil and duty to keep post-industrial humankind happy. But how much, and what kind? That is not a question that Siri or Alexa, or even Robot Radius, can answer.

Why 2020 was Twitter users' most miserable year yet

For more than a decade, researchers at the University of Vermont have been scraping a random sample of tweets posted on the microblogging site to gauge how happy or sad the world is feeling on any given day. Their tool, dubbed the Hedonometer, has been called the "Dow Jones index of happiness". It relies on a database of 10,000 frequently used words rated on a scale from 1 (least happy) to 9 (most happy). Every instance of the word "laughter", for example, receives a score of 8.5 out of 9; "happiness" and "love" rate almost as highly. At the other end of the scale, the words "suicide" and "terrorist" share the lowest score of 1.3. By averaging the "happiness score" of each day's tweets, the researchers can identify the events that have had the greatest impact on public sentiment around the world.

It should be no surprise that global happiness in 2020 was abysmally low. English-language Twitter's average score for the year was just 5.9, the lowest yet. It started on January 3rd with America's assassination by drone strike of Qassem Suleimani, head of Iran's Quds Force. As with the killing of Osama bin Laden in 2011, the Hedonometer swung sharply lower as Twitter posts were filled with negative words such as "died" and "killed". Sentiment tumbled again on March 11th when the World Health Organisation officially declared the covid-19 outbreak a pandemic. On May 31st, as protesters marched in more than 100 American cities after the police killing of George Floyd, users flooded the social-media platform with words such as "violence", "shooting" and "thugs", resulting in the saddest day in Twitter's history: the Hedonometer sank to 5.63. Although Western news tends to trigger the biggest movements in the Hedonometer's English-language happiness score, events elsewhere occasionally break through. When Nigerian security forces opened fire on peaceful protesters in Lagos on October 20th, the Hedonometer dropped to 5.78, just above the pre-2020 nadir of 5.75, which was recorded after America's deadliest mass shooting, in Las Vegas in October 2017.

Annus miserabilis

Average happiness score of English-language tweets

1=least happy, 9=most happy

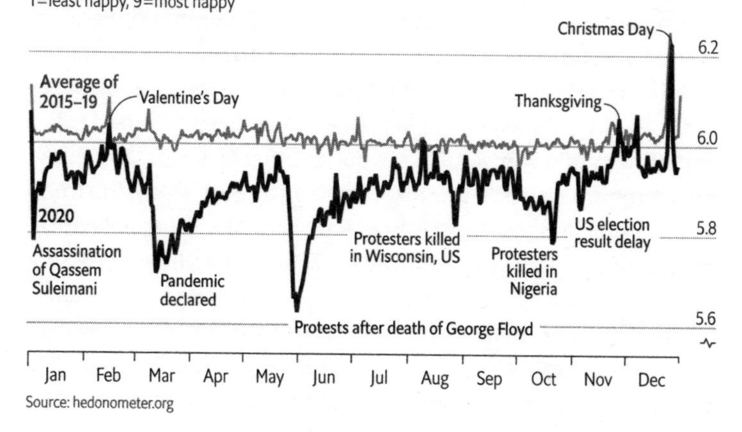

Source: hedonometer.org

The mood outside the English-speaking world was no less melancholy. The happiness score of French-speaking Twitter users averaged just 5.82 in 2020. The country suffered one of the most severe covid-19 outbreaks in the world and endured multiple lockdowns. Still, some countries were relatively cheery. South Korea weathered the pandemic well, and Korean tweets in 2020 featured many more happy words than French or indeed English ones did. It helps to have something to celebrate. Korean Twitter sees a huge surge in joyful words on the birthdays of the members of BTS, the world's most successful K-pop group. And the exuberance of their army of fans is spreading. When BTS scooped a number of music awards on December 6th, English-language Twitter recorded its second-highest happiness score of the year, buoyed by words like "winning" and "congratulations". Even in a gloom-filled year, happiness can be found on Twitter, if you know where to look.

Why Joe Biden's folksy speaking style is a strength, not a weakness

During the Obama years, the *Onion*, a satirical online newspaper, carried a series of articles about a man it referred to as "Diamond Joe" and "The President of Vice". It featured headlines like "Shirtless Biden Washes Trans Am in White House Driveway" and "Biden Huddling With Closest Advisers on Whether to Spend 200 Bucks on Scorpions Tickets". The joke was that the real Joe Biden, then vice-president, was a teetotalling and avuncular elder statesman, not a hair-metal-loving party animal. But the caricature of him as a regular guy drew on an element of truth. The son of a second-hand-car salesman, Joe Biden is the first president since Ronald Reagan to lack an Ivy League degree, having finished near the bottom of his class at both the University of Delaware and at Syracuse University's law school. This is reflected in his use of down-to-earth language.

In his speech, he is such an everyman as to defy parody. Forced to name Bidenisms, you might single out "malarkey", his favoured disparagement for humbug, or "Here's the deal", to signal that he's about to cut the malarkey. His favourite vocative is the folksy "Folks...". Mr Biden is, it is true, known for gaffes, though these are often overstated. While vice-president, he stood just a bit too close to a microphone as he said "This is a big fucking deal" into Barack Obama's ear at the signing ceremony for their health-care reform. He once asked a wheelchair-bound state senator to stand and be recognised.

These bloopers give the impression of a mouth running faster than a brain. Mr Obama was famous for long, thoughtful pauses; not so Mr Biden. His campaign speeches, like his questions as a longtime Senate committee chairman, tended to ramble. He once mystifyingly called a young student who asked a tough question a "lying, dog-faced pony soldier". Accused of corruption by a voter in Iowa, a bit of hotheaded "Diamond Joe" came out: he called the man a "damn liar" and challenged him to a press-up contest. And he mused that, were they in high school, he would "beat the hell" out of Donald Trump. Instead, he beat him in politics, with words rather than fists. The contrasting styles of his predecessors may have made Mr Biden's deeply prosaic register an asset after all.

Receiving his party's nomination in 2008, Mr Obama said future generations might remember the occasion as "the moment when the rise of the oceans began to slow and our planet began to heal". His high-flown rhetoric raised expectations to messianic levels. Dashed hopes led many voters to look for someone who sounded as little like a politician as possible. In 2016 that meant a political novice who eschewed focus-grouped formulations in favour of provocative, often vulgar tirades. Not only did voters not mind; Mr Trump's outrageous style was hugely effective. His vernacular delivery implied that he was a real boss, not a backslapping hack, with unique skills to get things done. But in office his coarseness turned some voters off, even as it thrilled others.

All that makes this an ideal time for Average Joe, for whom being able to talk fluidly at all was a hard-won achievement. Mr Biden grew up with a severe stutter, which he overcame as a young man. In one of the most touching scenes of his campaign, he told a boy who stutters, "Don't let it define you. You are smart as hell." Mr Biden took the boy's number and called him with some tips that had helped him; later the boy spoke to the Democratic National Convention. Even careful presidents make gaffes under constant scrutiny. But if Mr Biden gets tongue-tied or says the wrong thing every now and then, well, so do most people. After 12 years of extraordinary political speech, in one way or another, Americans were ready for a president who sounded like them.

Do languages evolve in the same way as biological species?

"Because politics." "Latinx." "Doomscrolling." Language is developing all the time, as new usages like these arise and old ones disappear. A common way to describe this process is to say that "language evolves". It is an apt formulation, for there is a deep and revealing relationship between linguistic change and biological evolution. But there are also some big differences.

Linguists today aim to apply methods from other sciences to messy social phenomena. But the influence once ran the other way, with discoveries in linguistic history leaving a mark on evolutionary theory. In the late 18th century William Jones, a British judge in Calcutta, concluded that Sanskrit's similarity to Latin and Greek was too great to attribute to mere chance. He proposed a parent language, the descendants of which included Sanskrit, Greek, Latin, Persian and many European tongues. Like Columbus, he was not the first to get there, but he made the revelation famous.

As Jones's findings were elaborated by the philologists who came after him, they also came to the attention of a young Charles Darwin. As early as 1837, looking at the evidence that wildly different languages had once diverged from a single parent, he wrote to his sister that mankind must have been around much longer than the Bible allowed. In 1871 he made the parallel between language divergence and evolution more specific, writing in *The Descent of Man* that "the formation of different languages and of distinct species, and the proofs that both have been developed through a gradual process, are curiously the same". One language giving birth to both Hindi and English was not so extraordinary if you gave tiny changes time to accrete.

Speciation – the emergence of distinct species – offers one of the closest parallels between linguistic and biological evolution. Darwin found that finches separated on different Galapagos islands had developed into different species, and worked out why. When a homogenous population is split, each subset will be affected by its own genetic changes. Those that contribute, even a little bit, to survival will tend to become more prevalent through the process of natural selection. When such changes accumulate, you no longer have two populations of a single species, but two different species.

Two linguistic populations separated by enough distance, or by a physical barrier such as a mountain range, can undergo a similar experience. Random alterations – to pronunciation, the meaning of words or grammar – are often so small that no one notices them as they are happening. Over the course of many generations, for instance, a *t* sound might become an *s*. Or take the terms in the opening lines of this article: using "because" as a preposition, shedding grammatical gender (as "Latinx" purports to do) or forging new words from old pieces (as in "doomscrolling") may all baffle the uninitiated. As tweaks of these and other kinds mount up in one group, its speakers gradually lose the ability to converse with another – as two speciating populations begin to lose the ability to mate. Mark Pagel of Reading University has made a list of other compelling parallels between the two processes. Like genes, he notes, words are "discrete, heritable units". The replication of DNA is akin to language teaching. Physical fossils resemble ancient texts. And so on.

But there are contrasts, too. Perhaps the biggest is that the chief driver of biological evolution – natural selection – is mostly absent in language. Nature is red in tooth and claw: a maladaptive mutation can get you killed. Language doesn't quite work that way. For the most part, changes don't take hold because they help you avoid a predator, but because they help people communicate. For that, they have to be adopted by others at the same time – which may happen for reasons that have little to do with "fitness". A celebrity's coinages will take off quicker than those of a brilliant basement neologist not because they are superior, but because the star has more Twitter followers.

There is, though, a final, important overlap between the two kinds of evolution. In a common visual depiction of the ascent of man, an ape gradually becomes a human through a series of intermediate steps. That gives the impression that evolution is a process of ever-increasing sophistication. Not always: rather, organisms, like languages, change to fit their environments. They may not always become more refined. But neither – despite the incessant chorus of grumbles – are they in decline.

Is "irregardless" really a word?

What, the actor Jamie Lee Curtis asked on Twitter on July 6th 2020, could make a dismal year even worse? Her somewhat surprising answer: "Merriam-Webster just officially recognised 'irregardless' as a word." A horrified emoji followed; 27,000 people signalled agreement with a "Like". Many others tweeted about their own dismay at this news. Yet the premise of these gripes was mistaken. Merriam-Webster had not "just" recognised *irregardless*. It had in fact appeared in Webster's controversial Third New International Dictionary of 1961, which also included words such as *ain't* and *heighth*, to the distress of many. It even made it into Webster's Second in 1934, which many linguistic conservatives cherish to this day as a totem from an earlier, more sensible age. *Irregardless* also turns up in the American Heritage Dictionary, which was explicitly created as a traditionalist response to Webster's Third, as well as in the capacious Oxford English Dictionary. What is it about *irregardless* that gets some people quite so worked up?

The case against it is straightforward. It has two negative affixes, one at the beginning (ir-) and one at the end (-less), making it malformed, those two negatives possibly even suggesting a positive. It probably began as an unintentional mash-up of *irrespective* and *regardless*. Does that mean it is not a word, though? Who would determine that, and how? All the words in this sentence, English-speakers would agree, are words. They can be found in dictionaries, and people know how to use them. And everyone would concur that an unpronounceable, random string of characters like *qtt3pf* is not a word. What about *klorf*? Harder to tell. (It's not a word: this correspondent just made it up. But unlike *qtt3pf*, it is pronounceable, so it could be.)

How about *hangry*? *Meh*? Real words? Slang? Half-words? It turns out that besides all the unambiguous English words, many others constitute a kind of penumbra of the language. These include dialect (such as Yorkshire's *nowt*, for "nothing"), terms from foreign languages being used increasingly in English (such as the Arabic word *hijab*), proper nouns that have become ordinary words (to *Google*), relatively new arrivals (*woke*), and others. Some dictionaries set out to make a mark by incorporating as many of these as they can – and their publishers try to drum up interest by putting out press releases touting the new entries. After all, weight aside, the most helpful dictionary is the one with the most words, not the fewest.

The case for admitting *irregardless* is not that it is lovely, or useful. It is simply that people sometimes say it. Lexicographers don't decide who gets into the club; they register who's already in, based on whether a word is in circulation. *Irregardless* has a fixed form (spelling and pronunciation) as well as a clear meaning: it's an adverb, used in the same way as *regardless*. But lexicographers are not unmindful of the standing of *irregardless* among the literati, so all the dictionaries mentioned here mark it as "nonstandard", or suchlike.

Even so, *irregardless* isn't nonstandard in the way that, say, *ain't* is. *Ain't* is extremely common, found in fiction, jocular standard speech and many dialects. *Irregardless* is much more of a fringe phenomenon. It hardly ever makes it into edited writing. Data from Google Books shows it to be only around 1/1,000th as common as *regardless* (and several of its rare appearances in print are in discussions about the word itself, like this one). The Corpus of Contemporary American English, another big database representing a wide swathe of language from different genres, finds it frequently in blog comments and unscripted television, but hardly anywhere else.

In other words, there is a better case against *irregardless* than the fact that it is malformed. (After all, many words are malformed: *television* combines Greek and Latin roots. *Flammable* arose from a misunderstanding of *inflammable*, which means "capable of being inflamed" but was misinterpreted as "not flammable".) The real problem is that it has never quite found a secure place in the English language. It may be said now and then by people who either do not think of it as an error or do not mind, but the fact that it hardly ever appears in edited prose – except when discussed as a solecism – suggests that it may never settle in. Perhaps another sense needs to be added to those dictionary definitions: "*irregardless*: a word that distinguishes people who do not care much about English usage from those who care terribly – and want the world to know it."

What wine vocabulary reveals about the nature of language

"Oak" and "fruit forward" are for wine amateurs. "Cedar" and "barnyard" are for real connoisseurs, and only a professional would have the confidence to deploy "gravel" or "tennis balls". One tasting note says a wine has hints of "mélisse, lemon balm". If you are wondering what "mélisse" is, don't bother: it is actually just French for "lemon balm". The language of wine is easy to mock. It can be recondite, even downright obscure. Oenophiles make a convenient subject for ridicule: if their cellars require such a wide-ranging lexicon, they are probably rich enough to cope with it. But wine vocabulary has its uses. Among the vast array of tastes, perhaps even flowery labels help experts pinpoint odours and flavours that they are interested in and want to remember. If you have a name for something, it may be easier to keep it in your head.

Perhaps. You might have heard the stereotyping joke about women having hundreds of words for colour in their vocabularies because they love to shop, but men having just the eight that come in a child's crayon box. This is a caricatured and simplified version of Ludwig Wittgenstein's view that "the limits of my language mean the limits of my world". The underlying argument is that having a name for something lets you understand it. But researchers have found that the links between perception, cognition and language turn out to be more complicated than that.

The debate over the relationship between thought and language is one of the most heated in psychology and linguistics. In one corner is the "Sapir–Whorf hypothesis", named after two early-20th-century American linguists, who posited that the world is made up of continuous realities (colour is a classic example) that are chopped into discrete categories by language. People thus perceive what their vocabulary prompts them to. An extreme version of this theory holds that it would be difficult, even impossible to distinguish colours – or wine odours or flavours – without names for them. On the other side of the debate are those who say that although language is indeed linked with cognition, it derives from thought, rather than preceding it. You can certainly think about things that you have no labels for, they point out, or you would be unable to learn new words. Supposedly "untranslatable" words from other tongues – which seem to suggest that without the right language, comprehension is impossible – are not really inscrutable; they can usually be explained in longer expressions. One-word labels are not the sole way to grasp things.

Into this dispute comes a study of wine experts and their mental labels. Ilja Croijmans and Asifa Majid of the Centre for Language Studies at Radboud University, in the Netherlands, gave a host of wine experts and amateurs a number of wines and wine-related flavours (such as vanilla) to sniff. Some from each group were told to name the odours they encountered; others were not. Then they were given a distraction to clear their minds, followed by a chance to recollect what they had smelled. As expected, the experts performed better than the amateurs – but those who articulated their thoughts did no better than those who had not. However, some who did not label the odours out loud may have done so in their heads. So the researchers conducted a second experiment. Some subjects were distracted while sniffing by a requirement to memorise a series of numbers, making it harder for them to verbalise what they smelled, even mentally. They did no better or worse than a second group who were given a visual distraction (memorising a spatial pattern), or a control group with no distractions.

The researchers conclude that olfactory memory in wine experts, at least, is not directly mediated by language. This is not to say such language is useless. Oenologists describe wines more consistently than amateurs do, meaning that – contrary to sceptical gibes about their pretentiousness – they are not just making up what they taste. Ms Majid says that rather than asking whether language affects cognition – since it clearly seems to, at least some of the time – the real question is what functions it affects. Perception, discrimination and memory are not the same thing, and some might be swayed by language more than others. Mr Croijmans compares words to a spotlight, which helps you separate things from the background, rather than giving you the ability to perceive them in the first place. That is a rather more positive version of Wittgenstein's aphorism: language not as a limit, but as a light.

Why Wikipedia's future lies in poorer countries

The mission statement "to organise the world's information and make it universally accessible and useful" belongs to Google. However, the search engine's ask-me-anything usefulness relies partly on the unpaid labour of the army of volunteer editors at Wikipedia, a collaborative online encyclopedia. Overseen by a not-for-profit group and dependent on constant donations of both money and time, Wikipedia is a brittle foundation for the world's informational ecosystem. The number of people actively editing Wikipedia articles in English, its most-used language, peaked in 2007 at 53,000, before starting a decade-long decline. That trend spawned fears that the site would atrophy into irrelevance. Fortunately for Wikipedia's millions of readers, the bleeding has stopped: since 2015 there have been around 32,000 active English-language editors. This stabilising trend is similar for other languages of European origin.

Meanwhile, as more people in poorer countries gain internet access, Wikipedia is becoming a truly global resource. The encyclopedia's sub-sites are organised by language, not by nationality. However, you can estimate the typical wealth of speakers of each language by averaging the GDP per head of the countries they live in, weighted by the number of speakers in each country. (For Portuguese, this would be 80% Brazil, 5% Portugal and 15% other countries; for Icelandic, it is almost entirely Iceland.)

Using this method, the richest Wikipedias – as measured by the share of speakers of each language who are active editors – tend to cluster in rich Western countries. For example, Wikipedia in Swedish, which is spoken by some 10m people, is managed by around 600 people a month. In contrast, Burmese, spoken by roughly 50m people, usually has fewer than a dozen people minding its site in a given month. This leaves the Wikipedias of most of the languages of Asia and Africa either bereft of articles or at the mercy of automation. Such sites are prone to including articles written by bots. After English, the language with the most

Wikipedia

Active editors* per million speakers v GDP per person† May 2015

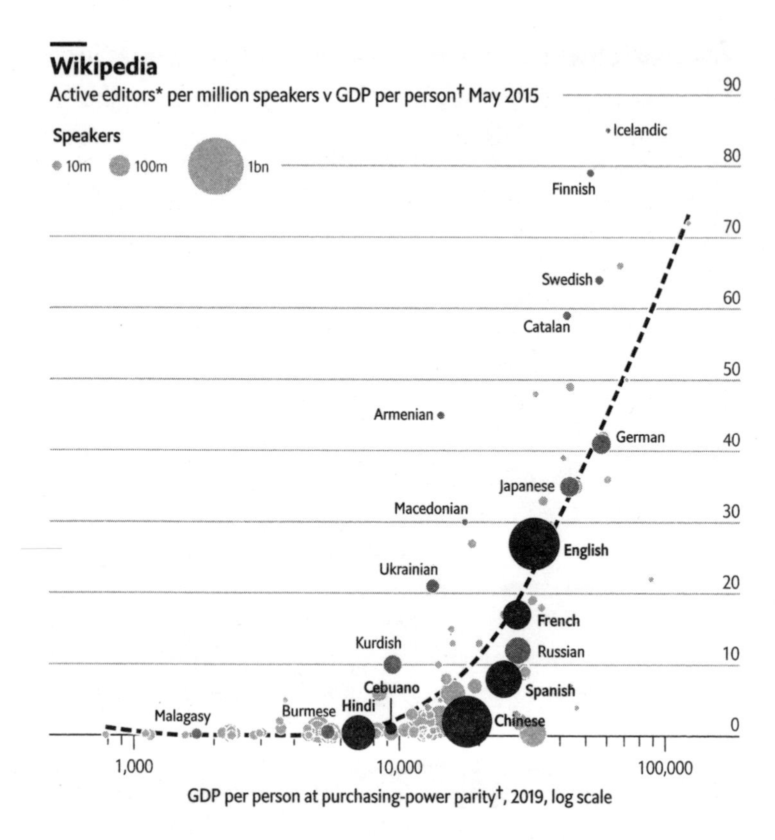

Total Wikipedia articles, m

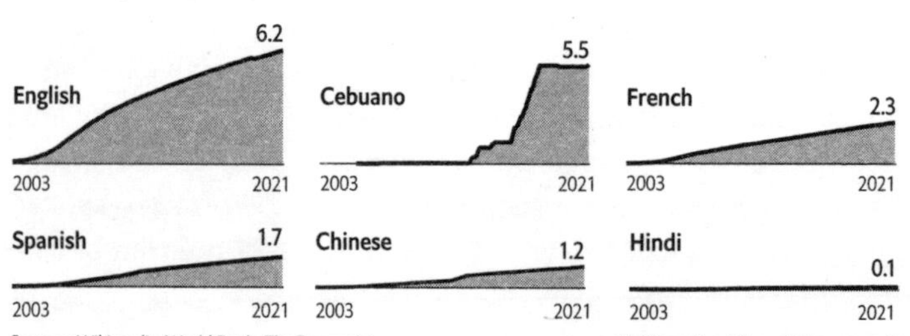

Sources: Wikimedia; World Bank; *The Economist*

*Editing at least five articles per month

†Average of countries in which speakers of each language live, weighted by number of speakers in each country

articles on Wikipedia is Cebuano, spoken by just 20m people in the Philippines. Nearly all were translated from English by a computer program created by a physicist in Sweden.

Users frustrated by clunky machine-written prose can soon expect a reprieve. Between 2010 and 2018 the number of active editors working in languages spoken in the richer half of countries in the world fell by 5%, but the corresponding figure for those spoken in the poorer half more than doubled. Wikipedia may have done the bulk of its organisation of the world's information long ago, but most of the work towards making it universally accessible and useful still lies ahead.

Why is the definition of genocide controversial?

The dictionary definition of "genocide" is simple. Just as "homicide" means killing a person and "patricide" means killing your father, so genocide means killing a people, such as an ethnic or religious group. The examples that spring most readily to mind are the Holocaust and, perhaps, the mass murder of Tutsis in Rwanda in 1994. As a result many people were perplexed when Donald Trump's secretary of state, Mike Pompeo, on his last full day in office in January 2021, used the word "genocide" to describe what China's government is doing to the Uyghurs, a mostly Muslim ethnic group, in Xinjiang, a western region of China. His successor, Anthony Blinken, agreed with him, but to many it sounded like the wrong word. China is undoubtedly treating the Uyghurs with horrific cruelty. It has locked up 1m or more of them in re-education camps, where they are beaten if they seem to revere Allah more than the president, Xi Jinping. But no one thinks China is carrying out mass slaughter in Xinjiang. So can its actions be defined as genocide?

The confusion arises because the UN's convention on genocide, which was drafted after the second world war, defines it exceptionally broadly, in ways that are quite different from the popular understanding of the term. It is not only killing that counts, according to the convention. So do "measures intended to prevent births", if their aim is "to destroy, in whole or in part, a national, ethnical, racial or religious group". So does "serious bodily or mental harm", if inflicted with the same aim, or the forcible transfer of children to a different group. Mr Pompeo cited reports of forcible sterilisation of Uyghur women to explain his use of the word "genocide". A Chinese official dismissed him as "a laughing-stock and a clown".

Defining genocide so broadly creates problems. If one reads the convention literally – and that is how one is supposed to read laws – an awful lot of governments can be accused of it. Officials in most countries have sometimes inflicted "serious bodily or mental harm" on members of ethnic minorities. How many people must they harm for it to count as trying to destroy that group "in part"? A million? A hundred? Ten? If too many crimes are classified "genocide", the term will start to lose its unique power to shock and shame. That power was already eroded by the political compromises that went into the drafting of the convention. The mass murder of class enemies was pointedly excluded – otherwise, Stalin would never have signed up. In practice, governments have long been reluctant to take the convention literally. America's State Department, for example, has previously reserved the label of "genocide" for acts that fit the dictionary definition, such as the mass slaughter of Muslims in Bosnia in the 1990s or of Yazidis in Iraq by the jihadists of Islamic State.

Some human-rights campaigners argue that calling the atrocities in Xinjiang "genocide" will stoke useful outrage and rally the world to oppose them. Others retort that making an accusation that the dictionary makes clear is false undermines the credibility of the accuser – and bolsters the Chinese government's argument that Westerners lie about Xinjiang to tarnish a rising power's reputation. Some countries prefer the phrase "crimes against humanity" to describe the persecution of the Uyghurs. That charge has the advantage of being obviously true – and therefore much harder for China to dismiss.

Why is America's best journalism published at the end of the year?

The winners of the Pulitzer prizes, the most prestigious in American journalism, are usually announced in April each year. Established more than a century ago by Joseph Pulitzer, a newspaper publisher, the awards are given for 15 categories of journalism – such as investigative reporting, feature writing and public service. There are also seven others, for letters, drama and music.

The 2020 winners, announced a few weeks later than usual, in May, were not surprising. The *New York Times* won three; the *New Yorker* took home two. The timing of the stories was fairly predictable, too. Pulitzer-winning articles tend to cluster around the end of the year, near when submissions are due (the deadline was recently moved from December 31st to the fourth Friday in January). Between 2014 and 2018, nearly a quarter of winners and finalists – excluding breaking news, cartoons and photography – hit newsstands and home-pages in December. In 2019 18% were published in the last month of the year.

Journalism is not the only field in which this sort of thing

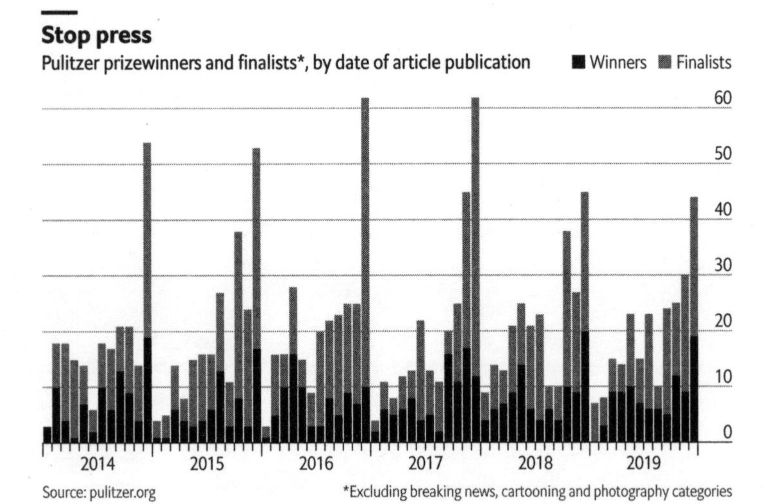

occurs. To be considered for the Academy Awards, or Oscars, films must have opened on big screens by New Year's Eve. Every year, a glut of hopeful contenders is released in December. The hope among film-makers is that their movies will be fresh in voters' minds as they cast their ballots, a phenomenon known as "recency bias". Sure enough, in recent years most of the films that have taken home the "big four" Academy Awards – best picture, director, actor and actress – were released in the fourth quarter of the year.

Are journalists, like film-makers, strategically timing their work? Perhaps. Or maybe recency bias works in a slightly different way in the news industry. Prizewinning publications might argue that they publish Pulitzer-worthy pieces all year round – and that their most recent projects, being the freshest, are just the likeliest to be submitted.

Facts of the matter: science, nature and the environment

Why Lithuanians suddenly embraced recycling

After the season for giving, 'tis the season for throwing things away. Each year in late December and early January a massive amount of plastic packaging is discarded worldwide. In Britain alone households generate 30% more waste, an extra 3m tonnes, in the month over Christmas. Most is destined for landfill. Lithuania does less damage than many countries, though. Its people now recycle things at a record level. Almost three-quarters (74%) of plastic packaging waste was recycled there in 2017, the highest proportion in Europe. The EU average was 42%, and the worst performers, including Finland and France, under 30%.

Much of Lithuania's success is due to a deposit refund scheme, introduced in 2016. Customers pay €0.10 ($0.12) extra when buying drinks containers. After use, these containers can be fed into "reverse vending" machines installed in shops, which spit the deposit back out. The machines' contents are then sent directly to recycling centres. By the end of 2017, 92% of all bottles and cans sold in Lithuania were being returned, close to triple the amount before the scheme began. The overall plastic packaging recycling rate increased by almost 20%. The Lithuanian government says the scheme has ignited a latent love for recycling in its citizens. Nearly 90% of Lithuanians have used the machines at least once.

However, Lithuanians do not generally describe themselves as eco-warriors. A 2017 survey by the European Commission found they were less likely than the citizens of most EU countries to regard environmental issues as "very important". The eagerness of Lithuanian recyclers may stem not from a love of the Earth but from a low net worth. Lithuania has the second-most unequal income distribution in Europe. A tenth of the population get by on less than €245 a month. In big cities it is common to see people scooping recyclable items out of bins in order to take them to the machines. There is a historical precedent, too: in the Soviet Union, bottle collection was often full-time work for those without other jobs.

Less litter and more money for people who need it seems like a win-win. But it might not in fact be best for the environment in the long run. In Germany – where a similar, widely used refund deposit scheme has been in place since 2003 – the earnings from keeping the deposits from unreturned bottles seem to have discouraged producers from switching to more sustainable packaging.

Why it matters that the Laptev Sea is freezing later

The Laptev Sea, nestled between the north coast of Siberia and Russia's New Siberian Islands, is known as the birthplace of ice. Every year, ice forms along its coasts in the autumn before being pushed west across the Arctic Ocean to Greenland and Norway's Svalbard archipelago, where it breaks up in the spring. Every year, that is, until 2020. For the first time since records began, the Laptev Sea had not begun to freeze by late October. This was only the latest example of how climate change is rapidly transforming the Arctic. Scientists worry that it could have damaging knock-on effects across the region.

The delayed freeze was the result of an unprecedented heatwave in Siberia in the first half of the year. The record high temperatures, in addition to fuelling an outbreak of Arctic wildfires that burned an area the size of Greece, warmed ocean waters by more than 5°C. That heat had not dissipated by late October, preventing the Laptev Sea from freezing on schedule. The peculiar weather was no coincidence. According to the World Weather Attribution project, a team of scientists that analyses possible links between climate change and extreme weather events, Siberia's scorching temperatures would not have been possible without climate change. Changes to Atlantic currents, also the result of climate change, may also have contributed to the lack of ice formation.

The Laptev Sea's woes are part of a broader transformation in the region. In the past four decades, the amount of so-called multi-year ice – sea ice that survives the warmer months – in the Arctic has fallen by half. Scientists expect that soon there may be none at all. What is happening in the Laptev Sea is a grim foreshadowing of this. In the 1980s, multi-year ice covered two-thirds of the sea; these days, ice can be absent for months at a time. This causes ocean temperatures to rise further, because open water retains more heat than ice. It also plays havoc with the Arctic's delicate ecosystem and is likely to threaten Arctic plankton, which play an important role

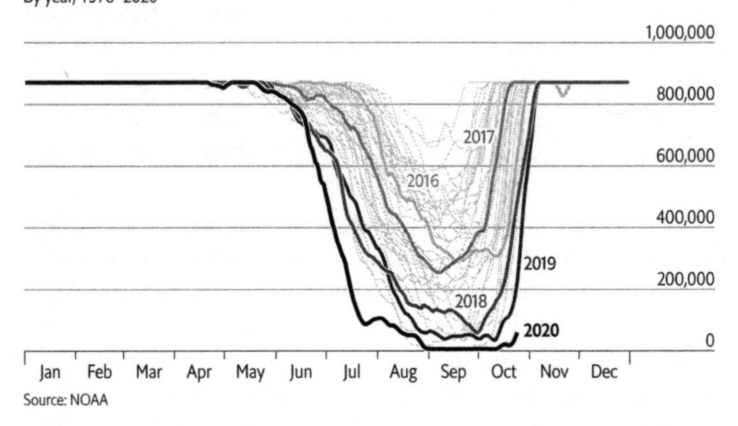

in absorbing carbon dioxide. The open water of the Laptev Sea is yet another warning of the consequences of a fast-warming world.

Why green investing has so little impact

Interest in climate change was once a rarity in high finance – it was the preserve of boutique investment houses and pokey back-offices in the large asset-management firms. Now it is all the rage. Pressure from regulators and clients, as well as the increasing frequency of extreme weather events, has made green investment red hot. The trend could be a force for good in the fight to reduce climate change. But there is a limit as to how much it can do.

In order to see how much of the world's emissions might be amenable to investor-led action *The Economist* analysed emissions disclosures from more than 5,000 publicly listed companies. The number of companies making such disclosures has been rising steadily. Those disclosures differentiate between the emissions that companies make directly, called "scope-one" emissions, and "scope-two" emissions which are produced by the companies which provide them with energy, mostly in the form of electricity. To look at the total emissions we considered only scope one, since adding in scope two leads to double-counting.

As you would expect, the largest emissions come from companies that burn fossil fuels in the normal course of their business: those that run fossil-fuel power stations, fleets of aircraft or steelworks. In Europe, ArcelorMittal is the biggest emitter because steelmaking requires the burning of coal. In America the biggest is ExxonMobil, which unlike many large companies produces much of the electricity and heat that it uses itself. Using the emissions disclosed by these companies, we estimated emissions for non-disclosing firms on the basis of those disclosed by similar firms in the same sector with comparable revenues. (Given that a firm's decision whether to disclose and its emissions intensity may not be independent, this process could lead to undercounting.)

Totting everything up reveals that each year publicly traded companies emit greenhouse gases equivalent to 10bn tonnes of carbon dioxide from their operations. Perhaps a quarter of those are produced by listed firms that are majority-owned by governments.

That leaves eight gigatonnes of emissions that stockmarkets can influence directly. That is only 14% of the world's total emissions, or 19% of emissions related to energy use and industrial processes. Fund managers therefore have some influence over a big slice of the economy, but many emissions occur outside the firms they control. They cannot directly influence the bosses of state-controlled Chinese coal-fired power plants or Middle Eastern oil and gas producers, for example. All this is a reminder that the role that financial services can play in fighting climate change should not be misunderstood – or overstated.

Why more people are making it to the top of Mount Everest than ever

Scaling Mount Everest was once a feat reserved for only the bravest mountaineers. But these days even relatively inexperienced alpinists attempt the climb. This is in part because it carries less risk. According to a paper published in August 2020 in the journal *PLoS ONE*, just 1% of those scaling the world's tallest mountain, with its peak at 8,848 metres (29,029 feet), die in the attempt. The fraction of climbers who manage to reach the summit, moreover, has more than doubled. Between 2010 and 2019 two-thirds of first-time Everest climbers reached the summit, up from less than one-third in the 1990s. (The 2020 climbing season was cancelled because of the pandemic.)

Even elderly thrill-seekers now have a shot at reaching the top of the world. The researchers found that success rates have increased across all age groups. Twenty years ago, climbers in their 60s had just a one-in-eight chance of reaching Everest's summit; now the odds are closer to one in three. Before 2006 nobody in their 70s had attempted to reach the peak for the first time. Since then the average success rate for septuagenarians has been 21%. Indeed, more than half of all climbers who try to scale Everest are now over 40 years old.

The fortunes of first-time Everest climbers have improved in tandem with the rise of the commercial climbing business. In the 1990s experienced guides began leading groups of less experienced climbers who, hitherto, might have gone it alone. Today at least 24 companies offer the service. Enthusiasts with little climbing experience can shell out tens of thousands of dollars to have their hands held most of the way up. And more climbers means an easier route, because part of the $11,000 fee that climbers pay the Nepalese government to climb the mountain goes to local mountaineers, who chart a path of ladders and ropes to the top. Their Tibetan counterparts do the same on the northern, Chinese-controlled side.

Better weather forecasting has helped too. Travelling from

Getting better all the time
Mount Everest, first-time summit attempts

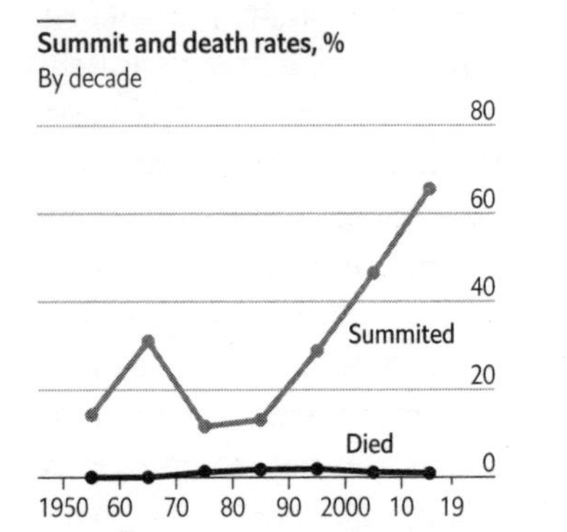

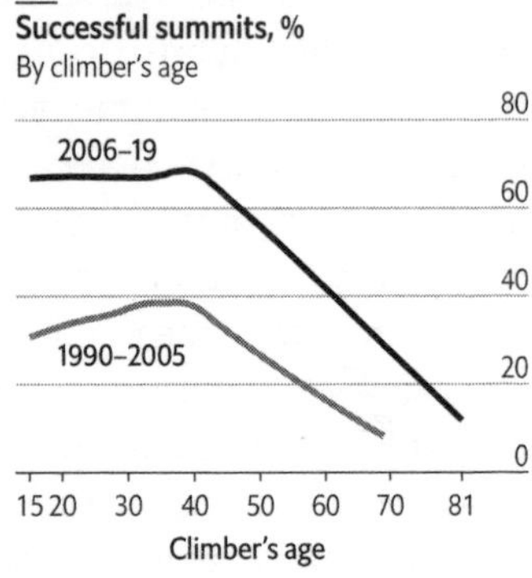

Source: "Mountaineers on Mount Everest: effects of age, sex, experience, and crowding on rates of success and death", R. B. Huey et al., *PLoS ONE*, 2020

base camp to the top of the mountain takes around four days, and success depends heavily on the weather along the way. High winds, caused by jet streams that can reach speeds of 120kph (70mph), and snowfalls are the biggest worries. Spotting a suitable "window" of good climbing weather in which to reach the summit was difficult 30 years ago. Today's high-tech forecasting techniques have made it much easier.

Some old hands fret that, now that the journey to the peak is easier, Everest is becoming overcrowded with inexperienced mountaineers. The 2019 season was a case in point: on May 23rd, 396 climbers tried to reach the summit in one day. The result was a long queue along a ridge in the "death zone" near the top, which may have contributed to the deaths of some of the 11 climbers who perished that month. It still pays to treat the world's highest peak with the utmost respect.

Why the age of forests, not just their area, is important

On August 6th 1964 an eager researcher sank a chainsaw into a gnarly Great Basin bristlecone pine on Wheeler Peak, a mountain in Nevada. A group of rangers then helped him cut and section the ancient trunk. Only later, when the young dendrochronologist had counted almost 5,000 years' worth of growth rings, did he realise that he had unwittingly deprived the world of its oldest known tree. Since then humans have been killing off the planet's old trees much more prosaically. A study published in May 2020 in the journal *Science* finds that the world lost more than a third of its primary forests (defined by the researchers as those undisturbed by humans for more than 140 years) between 1990 and 2015 to land-use change (eg, to create farms) and tree harvesting for wood. Over the same period, the area of younger forests almost tripled.

A team of more than 20 scientists analysed satellite imagery and 160 past studies to assess the factors affecting their death and regeneration of the world's forests. They found that remaining forests, even if they are not deliberately destroyed, are under stress from rising temperatures and higher carbon dioxide concentrations. This has led to a dramatic drop in the average age of trees – and, as a result, a reduction in their height.

Ancient woodlands are the arboreal sentinels of their ecosystems. They harbour more species than newer forests, thus preserving biodiversity. They are also the largest stores of carbon on land, thereby helping to mitigate global warming. The loss of primary forest in 2019 was associated with 1.8 gigatonnes of carbon-dioxide emissions, equivalent to that produced annually by 400m cars, according to Global Forest Watch, a monitoring service. As ancient trees die and decompose, heaps of stored carbon are released into the atmosphere. The vicious cycle begets more warming. Only some of this effect is offset by planting young trees to replace them, and suck up carbon as they grow.

Canopies have collectively become younger across the world.

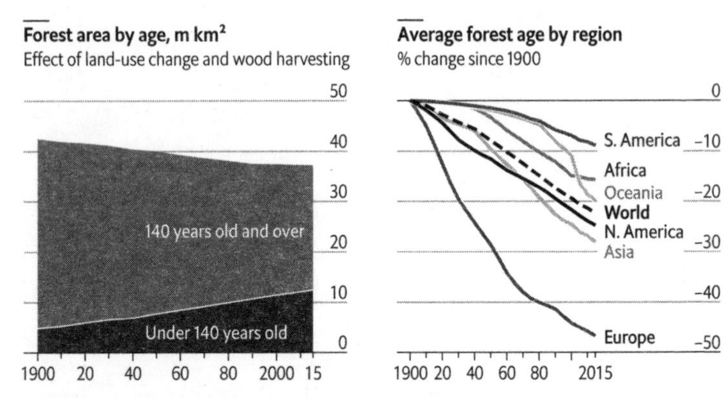

Source: "Pervasive shifts in forest dynamics in a changing world", N. G. McDowell et al., *Science*, 2020

The plunge has been sharpest in Europe, with the average age falling by almost half, while the smallest reduction has occurred in South America – of about 9%. This may seem counterintuitive, given the scale of deforestation in the Amazon. "Europe is declining fastest because it already started out with a very low average age in 1900, due to a long history of forest disturbance," says Louise Parsons Chini, a co-author of the paper. Europe is also bucking the deforestation trend, and planting new young woodlands lowers the average age in the mix. Between 1990 and 2015, Europe's forest cover has expanded by 90,000 square kilometres – an area as large as Portugal. By contrast, South America's deforestation has served to make room for cropland or pasture. The areas in question then drop out of the calculation for the average age of its forests.

Old trees are also dying at a faster rate than in the past. In North America and Europe, where the most data are available, tree-mortality rates have doubled in the past 40 years. Much of the loss is the consequence of decades of forest-harvesting, but rising global temperatures have also produced more wildfires, droughts and insect infestations. In recent years more fires have roared through Siberia, Australia and the Amazon. Droughts have grown longer and more severe. Scientists suspect that more frequent fires and droughts have also made trees less resilient to deadly insect invasions. For the first time, bark beetles have ravaged California's giant sequoias, the largest living organisms on Earth, despite their bug-repelling tannins.

Although higher levels of carbon dioxide are generally assumed to nourish tree growth, the study suggests this may be true only where water and nutrients are abundant. As dry periods become longer, water-stressed plants shed leaves and close up their pores to avoid moisture loss, thus limiting the intake of the greenhouse gas. "It's like an all-you-can-eat buffet, but with duct tape over your mouths," says Nate McDowell of the US Department of Energy's Pacific Northwest National Laboratory, who led the team behind the research published in *Science*. "Doesn't matter how much food there is if you cannot eat it." All this means the world's forests are, collectively, becoming smaller, younger, shorter – and more vulnerable.

Why dung beetles' love of human faeces results in scientific bias

Ecosystems are complex things, and monitoring their health is hard. To track every species would be impossible, so ecologists commonly focus on those that, like canaries in coal mines, are thought to indicate when the system as a whole is beginning to suffer. Dung beetles are one such group. They have been relied on heavily for years to monitor the effects of things like logging, grazing and road-building. When there are lots of different species of dung beetles around, and faeces vanish quickly, an ecosystem is assumed to be in good shape. When their diversity drops and faeces hang about unconsumed, it suggests something is wrong.

However, as Elizabeth Raine, a zoologist at the University of Oxford, has realised, the value of this assumption depends on how you sample the beetles. That is done by attracting them with their preferred foodstuff – faeces. And she thinks it is being done badly. Researchers had assumed that dung beetles would happily tuck into any old pile of dung. As such, they have a cherished tradition of using their own excreta as bait. This makes sense, because a supply is always available. But Dr Raine realised that no one had ever tested how attractive human faeces were compared with those of wild animals.

She and her colleagues therefore set up experiments at three lowland tropical-forest sites in Paraná, Brazil. They created pit traps around areas in which they had placed faecal lures. Rather than use human waste, they employed droppings collected from the enclosures of animals in a local zoo. These included lesser grisons (members of the weasel family), jaguars, ocelots (a species of small wild cat), crab-eating raccoons, domestic pigs and black capuchin monkeys – all species that inhabit the forests in question. For comparison, Dr Raine also used her own faeces in some sites, as a control.

The results, published in the journal *Biotropica*, suggest that – in lowland Brazilian forests, at least – dung beetles of all sorts are particularly fond of human waste, and much less interested in the dung of the animals native to their ecosystems. After 48 hours an average of 20 beetles were found in the pit traps next to human waste, whereas jaguar faeces lured an average of just ten and the dung of other species even fewer. Diversity of beetle species was also far higher in traps near human excrement than it was near the other samples.

This is both encouraging and problematic. It suggests that using human faeces as a lure is a good way to get a general sense of which types of dung beetle are present in an area. The drawback is that human faeces are so alluring to these insects that using them may be creating an illusion that they are present in greater numbers than is actually true, and consuming more wild-animal dung than they really are. Clearly, Dr Raine's experiment needs to be replicated in other habitats, to check that she has discovered a general phenomenon rather than one specific to a particular area. But if it turns out that most dung beetles do indeed prefer human faeces to their normal meals, it means that a lot of previous ecological studies may need to be reinterpreted.

Why Dr Raine's dung beetles prefer human faeces is a mystery, though she suspects that the varied diets enjoyed by modern human beings may have something to do with it. But her experiment does illustrate a wider point. It is that scientific discoveries are only as good as the experiments used to make them. It therefore behoves people to check even well-established procedures from time to time, to make sure they are not falsely assuming that what is sanctioned by familiarity actually works.

Why the Nobel prize delay is growing

Albert Einstein doubted the existence of black holes. In 1939 he even tried to prove that the celestial objects – which are so dense that light itself cannot escape their gravitational pull – do not exist in the "real world". But in 1965, ten years after Einstein's death, Roger Penrose, a British physicist, wrote a groundbreaking paper that used maths to prove that black holes are a necessary consequence of the theory of relativity. In 2020, aged 89, Sir Roger won the Nobel prize for physics for his seminal work. It may seem odd to award science's most prestigious prize for a discovery made 55 years ago. But the period between when Nobel-worthy discoveries are made, and when they are recognised, much like space-time itself, has curved upwards since the prizes were first handed out in 1901.

A long wait for Nobel recognition is nothing new. Peyton Rous, an American pathologist, was first nominated for the medicine prize in 1926, for his discovery in 1911 that cancer could be transmitted by a virus. He would have to wait until 1966 for the accolade. And some breakthroughs are recognised speedily: the 2020 chemistry laureates, Emmanuelle Charpentier and Jennifer Doudna, won for their work on gene-editing, which they did as recently as 2012.

But data compiled by researchers at the University of Florence, Aalto University in Finland and the University of Belgrade suggest that this "prize delay" now routinely spans decades. In physics, over the past century the average gap between achievement and award has increased from ten years to more than 30. Fifteen of the longest 20 gaps have come in the past two decades. These include British scientist Peter Higgs (2013) who was awarded the Nobel prize 49 years after predicting the existence of the boson that bears his name, and Nambu Yoichiro (2008), a Japanese scientist, who became a laureate 48 years after his work on spontaneous symmetry-breaking. In chemistry, the average delay has roughly doubled in the past 100 years. In medicine, the same trend is evident. The 2020 prize went to three scientists involved in identifying the hepatitis C virus – way back in 1989.

Better late than never

Nobel prizes, time lag between discovery and award, years

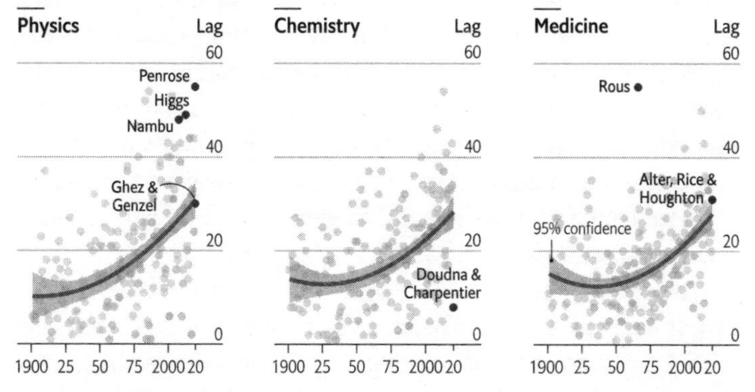

Source: "The Nobel Prize delay", F. Becattini et al., *Physics Today*, 2014; *The Economist*

There are many theories about why the lag is growing. Increased life expectancy may play a part. Scientific progress may also be slowing, forcing the awards' committees to look ever further into the past for deserving nominees. But perhaps the most convincing explanation is that Nobel prizes are awarded for discoveries, not unsupported theories, and complex ideas may take longer to prove than they did a century ago. After all, Sir Roger Penrose's mathematical breakthroughs could not be confirmed empirically without technological innovations developed during subsequent decades. Whatever the reason, today's scientists should hope that the curves start to flatten. Otherwise it is likely that their crowning moment of glory will see them with one foot in the grave.

How the shark forgot his skeleton

Vertebrates – animals with a backbone – are a diverse bunch, encompassing everything from tuna and budgerigars to snakes, chinchillas and human beings. One way that biologists divide them up is based on the composition of their skeletons. Most vertebrates sport hard, calcified bones, and are dubbed the osteichthyans. A second, much smaller category is the chondrichthyans, whose members include sharks, rays and skate. Unlike their hard-boned cousins, chondrichthyans make do with structural parts made of soft, tough cartilage.

Palaeontologists had long assumed that cartilage was the more primitive arrangement. Osteichthyan fetuses, after all, begin life with a cartilaginous skeleton that is gradually replaced by harder, more durable bone as they grow. But a paper published in *Nature Ecology & Evolution* suggests that this view may be mistaken. The paper describes the discovery of a 410m-year-old fossil in Turgen, a district of Mongolia close to the Russian border, by a team led by Martin Brazeau of Imperial College London. The fossil is the partial skull of a new species of placoderm, a type of armoured fish, which Dr Brazeau and his colleagues have dubbed *Minjinia turgenensis*.

Placoderms are of interest to palaeontologists because they are an immediate common ancestor of both the chondrichthyans and the osteichthyans, which are thought to have split from each other around 400m years ago. Yet when Dr Brazeau's team scanned their fossil with X-rays, they discovered tissues defined by microscopic curved struts and rods called trabecles – very similar to the bones found in modern-day bony vertebrates. If an ancestor of both the chondrichthyans and the osteichthyans had the ability to grow a hard skeleton, that implies that, rather than bony fish inventing the trick, it was the ancestors of today's sharks and their cousins that forgot it.

Exactly why is a matter of conjecture. One possibility is that it was an evolutionary adaptation. "Sharks don't have swim bladders, which evolved later in bony fish, but a lighter skeleton would have helped them be more mobile in the water and swim at different depths," speculates Dr Brazeau. This may be what helped sharks become one of the first global fish species, spreading out into oceans around the world 400m years ago, he suggests.

Whatever the reason, the fact that sharks are still around 400m years later suggests that a soft skeleton is a highly successful strategy, from an evolutionarily perspective. Yet it may also be a limited one. The swim bladders developed by bony fish to modulate their buoyancy would soon be co-opted for another purpose: transformed into a primitive set of lungs, they allowed their bearers to breathe air, and therefore to colonise the land. The result was terrestrial vertebrates. And the rest, as they say, is history.

How to spot dodgy academic journals

As covid-19 spread in early 2020, scientists raced to study it. Although scientific journals have tried to speed up the "peer review" process that they use to check new findings, many researchers bypass the process altogether by uploading unchecked working papers to preprint sites. This means that flimsy findings, in virology and other fields, can travel as fast as a novel coronavirus. Most scholars who share preprints are doing their best to make vital discoveries. However, some authors seek to pad thin résumés by publishing underwhelming, repetitive or fake research. As safeguards are relaxed, journalists and governments need to be on high alert to spot such studies.

Such results mostly appear in "predatory" journals, which make use of the popular "open-access" model – charging fees to authors, rather than to readers – to publish any old tosh for money. According to Cabells, a firm that maintains a blacklist of such journals in English, some 1,000 existed in 2010. By 2020 there were at least 13,000. Some scammers are careless. Mike Daube, a professor of public health, got his dog Ollie, a Staffordshire terrier, onto seven journals' boards, to expose their lax scrutiny of candidates. Cabells uses 65 criteria to spot wilier frauds. "Severe" infractions, such as missing back issues, lead straight to blacklisting. Lesser ones, like poor spelling or offers of speedy publication, set off further investigation.

The field covered by a journal offers few clues to its trustworthiness. The mix of topics is similar on Cabells' whitelist, of 16,000 reliable journals, and its blacklist. On both, a third of titles relate to health. Geography is more revealing. Cabells lists only a few reliable Nigerian journals, but 1,100 predatory ones. India's figures are 300 and 4,400 respectively. Another 5,800 blacklisted titles claim to be based in Europe or North America but do not provide evidence, such as a valid address. The authors of dodgy papers published in such journals are often from developing countries, but Western academics have been caught red-handed as well. Many scholars claim to have been duped into using such journals.

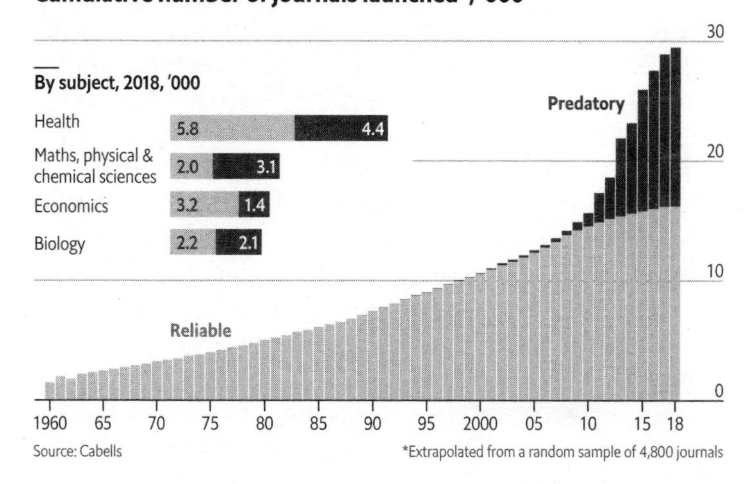

The average predatory journal publishes about 50 articles a year, less than half the output of a reliable title, according to Bo-Christer Björk of the Hanken School of Economics in Helsinki. And 60% of papers in such journals receive no future citations, compared with 10% of those in credible ones. Still, that leaves 250,000 questionable articles per year that do get cited. The fact that so many "scholars" publish results in dodgy journals means that working papers on covid-19, and on many other life-or-death topics, should face extra-thorough scrutiny.

Playing by the rules: sport, games and leisure

How empty stadiums made it possible to measure referees' bias

Coronavirus lockdowns were particularly frustrating for football devotees, who had no live matches to watch while stuck at home. But the fans most pleased by the sport's return may be statisticians. For them, empty stadiums are not a cheerless last resort, but rather a chance to tackle a great quandary: why do travelling teams tend to lose? Most studies have blamed referees for trying to placate the home team's fans. In one experiment, officials were shown recorded games and asked how they would have ruled. They were kinder to home sides when they could hear baying fans than when the sound was muted. Some analyses of live matches have found more bias with denser crowds.

Before the summer of 2020, few competitive fixtures were played without fans. One study from May 2020 found just 160 cases since 2002. In that small sample, the home team's edge vanished. Referees gave similar numbers of cards for fouls to both sides, and visitors won almost as often as hosts did. This finding could easily be skewed by a few clumsy tackles or lucky goals. But the pandemic made a larger study possible. *The Economist* asked 21st Club, a consultancy, to analyse 1,534 matches played without fans in 2020, spanning dozens of leagues.

Sure enough, officials no longer appear biased. Although the pattern varied widely by league, the total share of cards received by home teams rose from 46% before lockdowns to 50% afterwards. But the lack of help from referees merely reduced home sides' advantage, rather than eliminating it. Football leagues give teams three points for a win, one for a draw and none for a loss. With crowds watching, home teams gained 58% of points; without them, hosts still earned 56%. In other words, three-quarters of home overperformance remains intact.

So why do teams fare better at home, even without biased officials? Between May and July 2020, hosts took 53% of shots at goal – less than the 55% they took in full stadiums, but enough

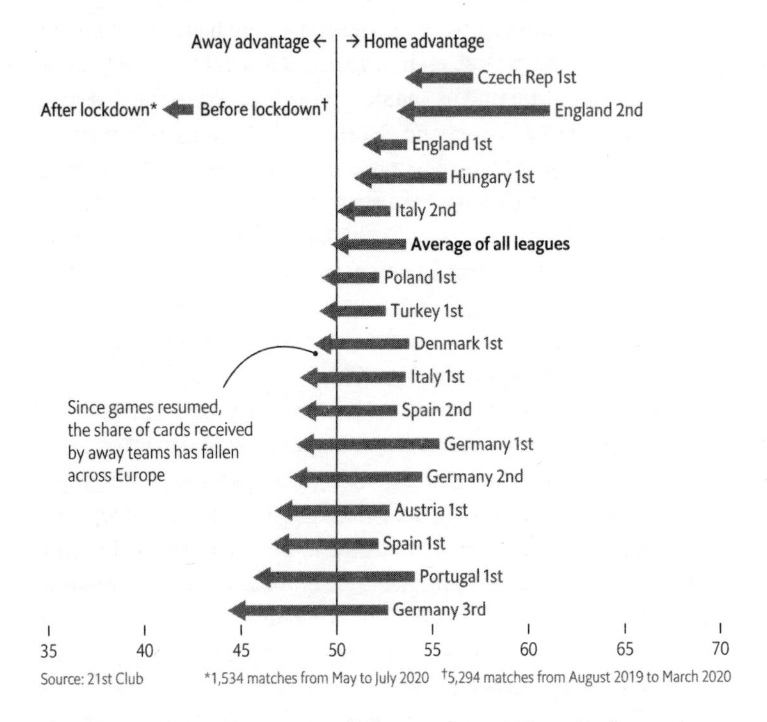

to give them an edge. The cause of this home-team advantage remains mysterious. Perhaps travelling to an away game affects performance. In American sports, visitors have to endure long trips between games. But travel times within European countries rarely exceed a few hours. A more plausible reason is that team coaches still use conservative line-ups and strategies when playing away, even though hostile officials no longer undermine their best players' efforts. If that is a factor, managers who excoriate referees ought to consider their own failings, too.

Why African countries issue stamps celebrating English cricketers

Len Hutton was an accomplished cricketer. English fans cherish the record 364 runs he racked up in a Test match against Australia in 1938. It would not be unreasonable to surmise, however, that this feat is less remarked on in the Central African Republic (CAR), a former French colony with no cricketing pedigree. So it may seem odd that in 2016 the CAR issued a set of commemorative stamps to mark the centenary of Hutton's birth – and odder still that French-speaking Niger and Portuguese-speaking Mozambique did the same.

The practice, it seems, is not restricted to bygone English cricketers. The 395th anniversary of the death of Jan Brueghel the Elder, a Flemish painter, occurred in January 2020, a milestone Sierra Leone's postal authorities considered significant enough to warrant a philatelic tribute. Other African states seem to prefer Baroque music to Flemish art. It is just possible that the people of Guinea-Bissau might have let the 260th anniversary of the death of Johann Sebastian Bach slip by without much fanfare. Fortunately, their postal service was less remiss. Countries normally celebrate national heroes with special stamps. Yet there is a rationale for printing stamps of long-deceased foreigners. You would be hard-pressed to walk out of a post office in Bangui, the CAR's benighted capital, with a Hutton stamp. But collectors scouring the internet are willing to pay handsomely for such stuff.

A set of Hutton stamps from the CAR fetches €15.50 ($18.50). Niger's depiction of a besuited Hutton is sufficiently sought-after to command a €2 premium. All this is a handy way for African states to boost revenues. Some philatelists get prissy about this sort of thing. Stamps should be issued only to meet local demand, they insist. Yet targeting the collectors' market is not new. Smaller and poorer countries have been at it for 70 years, notes Ian Harvey of the Royal Philatelic Society London. Even Britain's august Royal Mail is accused of collector-gouging.

But there are signs that the practice is getting out of hand. In recent years several African countries have appointed a Lithuanian-based outfit called Stamperija to design and print their stamps. Stamperija, philatelists grumble, has flooded the market with tat. Collectors calculate that, with the help of Stamperija, Sierra Leone, with a population of 7.6m and a barely functioning postal service, churned out 1,566 different stamps in 2019, compared with 268 released by Britain and 139 by India. Stamperija's gaudy designs are not to everyone's liking. And producing so many stamps can lead to mistakes. A set of Stamperija stamps issued in 2014 for the CAR turned out to picture not Marilyn Monroe, as planned, but a New York drag artist. Few of Stamperija's clients seem bothered, though. "It is willing buyer, willing seller," says a postal official in Sierra Leone. "So what's the problem?"

Why are penalties becoming more common in elite football?

In football, winning a penalty kick is a big step towards winning a match. Around 75% of penalties are successful. Teams that are awarded them go on to win or draw in more than 80% of matches, according to an analysis of World Cup and European Championship matches published in 2015. In the Premier League, the top tier of English football, penalties are on the rise. The highest number awarded in a season is 106, in 2009-10; this tally was matched in 2016-17. It seemed sure to be surpassed in 2020-21 which, only halfway through, had already notched up 71 spot-kicks.

There are three reasons for the increase. First, the rules that most often result in penalties are now being more strictly enforced. Tackling from behind was outlawed in the 1990s, and in 2019 the International Football Association Board (IFAB), which oversees the rules of the game, ruled that touching the ball with an arm or hand, even if accidentally, is a foul. Second, the introduction of the video-assistant referee (VAR), which allows on-field referees to consult replays of incidents, has helped officials spot penalty-worthy offences. Third, cultural perceptions have changed. The act of "diving" – falling to the ground to create the appearance of a foul – was once almost universally frowned upon. But today TV pundits routinely question players who choose to stay on their feet after getting roughed up, when they could hit the deck and earn a free-kick or penalty instead. Persuading referees to award fouls for marginal decisions has become an art in itself. Marcus Rashford, a striker at Manchester United, has spoken of how José Mourinho, a former manager of the club, taught his players how to become more "savvy" in winning penalties.

More teams may catch on to the benefits of such tactics. In the Premier League's 28 years, 11 teams have been awarded more than ten penalties in a season; eight of these occurrences have come in the past ten seasons. In the first half of the 2020-21 season, Leicester City was winning penalties at a rate of one every two games. In

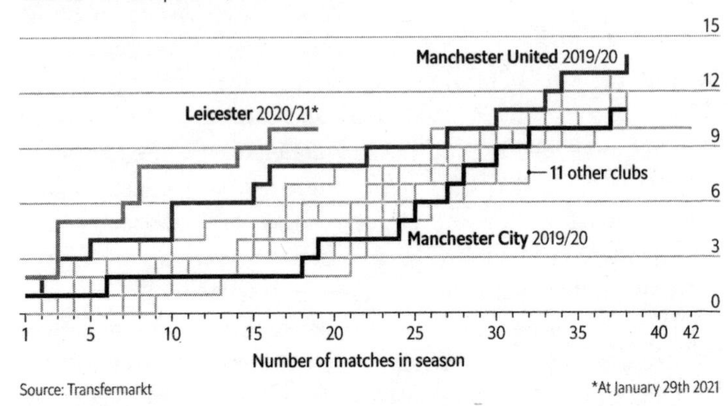

many ways, the team's strategy – which used lots of through-balls to strikers that can cause goalkeepers to concede penalties – resembled that of Crystal Palace in the mid-2000s and Leicester ten years later. Those two teams rank among the sport's most prolific winners of penalties.

How new swing techniques are revolutionising golf

Golf has a length problem. The farther players drive the ball, the longer holes need to be, so that skills like iron play and putting remain important. But the longer courses are, the more they cost to maintain and the worse their environmental impact. They also become more daunting for recreational golfers, who keep them in business. In 2004 golf's regulators introduced limits on the size of clubs, hoping to slow the trend of ever-longer drives. Nonetheless, the inflation continued apace. On November 15th 2020 a famous record tumbled: someone completed the Masters Tournament in fewer than 270 strokes, the mark Tiger Woods set when he won his first major title in 1997. The new low of 268 belonged to Dustin Johnson, a burly driver who has averaged more than 300 yards (274 metres) a pop throughout his career. He achieved the feat even though the Augusta National course is 8% longer than it was in 1997. How have golfers continued to blast the ball farther than ever?

The PGA Tour, the top men's circuit, publishes ball-tracking statistics based on the flight of most drives in tournaments since 2007. These suggest that, although better equipment may have helped, players' recent gains stem largely from their technique – and even bigger improvements now appear inevitable. The data come from ShotLink, a system that tracks how fast a golfer swings ("clubhead speed"), the ball's trajectory ("launch angle") and its rotation speed ("spin rate"). After taking each player's average value for these metrics in each year, we used them to build a statistical model to predict driving distances. Together, the three factors explained 70% of the differences between players' distances, and almost all of the increase in length over time.

The model's lessons are intuitive. To thump the ball as far as possible, maximise clubhead speed and launch angle while minimising spin (which causes the ball to soar higher, rather than racing forward). However, most players face a trade-off between these goals, explains Paul Wood of Ping, a club manufacturer.

Tee time

Predicted effect of launch angle on driving distance

With 135mph swing speed and 2,200 rpm spin rate

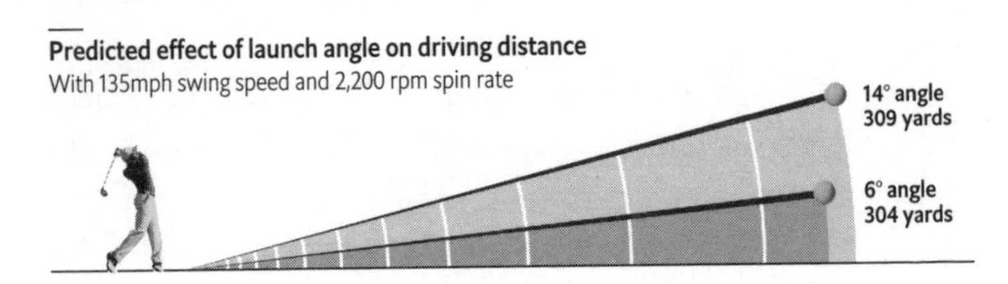

PGA Tour golfers, 2007–21 seasons*, average swing speed v launch angle of ball

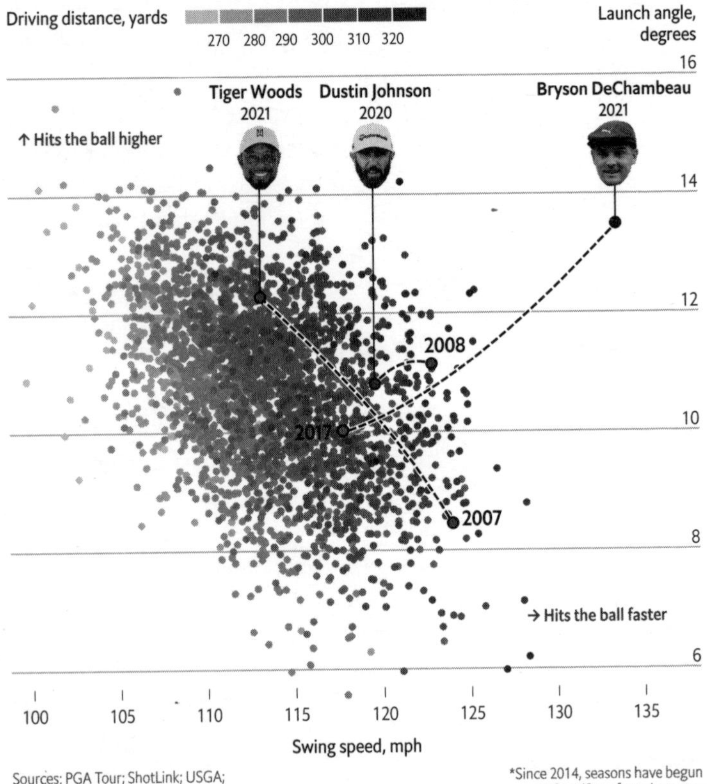

Sources: PGA Tour; ShotLink; USGA; R&A; Distance Insights Project

*Since 2014, seasons have begun in Sept/Oct of previous year

Harder impacts usually mean flatter trajectories. Although the average male player swings faster and produces less spin than in 2007, launch angles have declined since then.

One golfer, however, has escaped this constraint. Bryson DeChambeau, a physics graduate with oddly designed clubs and a voracious appetite for data, is nicknamed the "Mad Scientist". While the PGA Tour was suspended because of covid-19 in 2020, he added 18kg (40lbs) of bulk. This then allowed him to swing faster than anyone else. But he also managed to smash the ball with a high launch angle, rather than a low one – an unprecedented combination that might owe something to his unusually stiff wrists and robotic technique. Using both his brains and his brawn, Mr DeChambeau is now hitting 15 yards farther than his closest competitors do. He won his first major title at the US Open in September 2020.

Mark Broadie, a professor at Columbia University and a golf statistician, reckons that other professionals will try to beef up in response. But golf history is littered with players who lost their edge after tinkering with their swings. And time may yet show that the risks of Mr DeChambeau's bombs-away approach offset some of the rewards. He often strayed into the rough at the Masters. Nonetheless, the Mad Scientist's breakthrough is bad news for course designers. They will probably have to keep fiddling with their fairways for years to come.

Why are so many athletics records falling?

Elliot Giles, a previously unheralded British middle-distance runner, won the men's 800-metres race at an event in Poland in February 2021. He did so in style. His time – 1 minute, 43.63 seconds – was the second-fastest indoor 800 metres in history. He took more than a second off the British record, which had stood for almost 40 years. Mr Giles's feat was the latest in a succession of impressive performances in track athletics. In February 2020 a Scottish 800-metre runner, Jemma Reekie, knocked four seconds off her personal best and ran the 11th-fastest indoor time ever. In October 2020 an Ethiopian, Letesenbet Gidey, took more than four seconds off the women's world record in the 5,000 metres. An hour later a Ugandan, Joshua Cheptegei, reduced the men's 10,000-metre record by six seconds. What do all of these athletes have in common? They were wearing next-generation running shoes.

Shoemakers, most notably Nike, but also Adidas, Asics and New Balance, have chased records for years. Nike designed an event in 2017 in which long-distance athletes attempted to run a marathon in less than two hours using the company's Vaporfly shoes. (None succeeded, although Eliud Kipchoge, the reigning Olympic champion, eventually did so at an unofficial event in 2019.) The Vaporflys have since been shown to improve runners' efficiency by around 4%, which translates into a reduction in marathon times for elite male runners of around 90 seconds. These savings are generated through a soft and springy foam in the sole and a carbon-fibre plate, which can preserve runners' energy and reduce the loads on their bodies. (Scientists are still unsure exactly how these elements combine, and shoemakers are cagey with the details.) With clear evidence that road athletes are running faster, manufacturers have now turned their attention to spikes for the track.

A big part of the appeal of athletics is the prospect of tumbling records. But until recently the pace of record-breaking seemed to be slowing down. In the 1960s, seven men set new world records over 400 metres, reducing the time by almost 1.5 seconds over the decade. In the subsequent 50 years, the record has been broken only three times and by a cumulative 0.8 seconds. A similar stagnation is true in women's events: the world records in the 100, 200, 400 and 800 metres have all stood since the 1980s. (At the time the sport was dogged by doping accusations; the record-holders have denied cheating.) Without big stars who can improve on their predecessors, such as Usain Bolt, athletics struggles to excite anyone except existing fans.

The recent spate of records may seem a good thing, but it raises awkward questions for World Athletics, the body that regulates elite running. Short of performance-enhancing drugs, technological progress has always been permitted in the sport, from replacing cinder tracks with synthetic rubber to using digital watches to record times precisely. Yet it will be keen to avoid the kind of farce seen in swimming in 2008-09, when more than 100 world records were set in 18 months after advanced polyurethane suits were permitted. Swimming's governing body subsequently banned the suits, but many of the tainted records still stand.

The president of World Athletics, Sebastian Coe, whose British indoor record was broken by Mr Giles, has said the sport "should not be in the business of suffocating innovation". Guidelines introduced in 2019 permit almost all of the latest models of shoe to be used in major competitions, provided they are also available to the general public. Yet if one firm's shoes emerge as noticeably faster than the rest, athletes sponsored by other manufacturers will be at a disadvantage. Mr Giles, for his part, calls any suggestion that his performance was down to fancy footwear an insult. Meanwhile, some runners have taken to wearing disguised versions of Vaporfly. The controversy around such "super shoes", it seems, will run and run.

Why engineers, not racers, are the true drivers of success in motor sport

"I always thought records were there to be broken," Michael Schumacher, a star Formula 1 (F1) driver, said in 2013. At the time, his record of 91 career F1 victories looked safe: the closest active racer had just 32. Yet on October 11th 2020 Lewis Hamilton of Britain equalled the mark. The following month he went on to equal Mr Schumacher's record of seven F1 championships. Mr Hamilton's ascent has ignited debate over whether he is F1's best driver ever. Comparing athletes across eras is always hard – especially in motor sports, where a racer depends on his car. Moreover, F1 has regularly changed its scoring system and its number of races, drivers and teams.

However, statistical analysis can address many of these nuances. *The Economist* built a mathematical model, based on a study by Andrew Bell of the University of Sheffield, to compare all 745 drivers in F1 history. It finds that Mr Hamilton's best years fall just short of those of the all-time greats – but so do Mr Schumacher's.

The model first converts orders of finish into points, using the 1991-2002 system of ten points for a win and six for second place. It adjusts these scores for structural effects, such as the number and past performances of other drivers in the race. It then splits credit between drivers and their vehicles. Disentangling these factors is tricky. Mr Schumacher spent most of his peak at Ferrari, as Mr Hamilton has at Mercedes, leaving scant data on their work in other cars.

However, their teammates varied. (Today, F1 has ten teams, each using two drivers and one type of car.) And drivers who raced alongside Mr Hamilton or Mr Schumacher tended to fare far better in those stints than they did elsewhere. If Ferrari's and Mercedes' engineers boosted lesser racers this much, they probably aided their stars to a similar degree. Because most drivers switch teams a few times, this method can be applied throughout history.

Between the two racers with 91 wins, the model prefers Mr

Pole position

Relative importance of car quality to driver skill for champion driver

Standardised Formula 1 points*, expressed as % of maximum possible

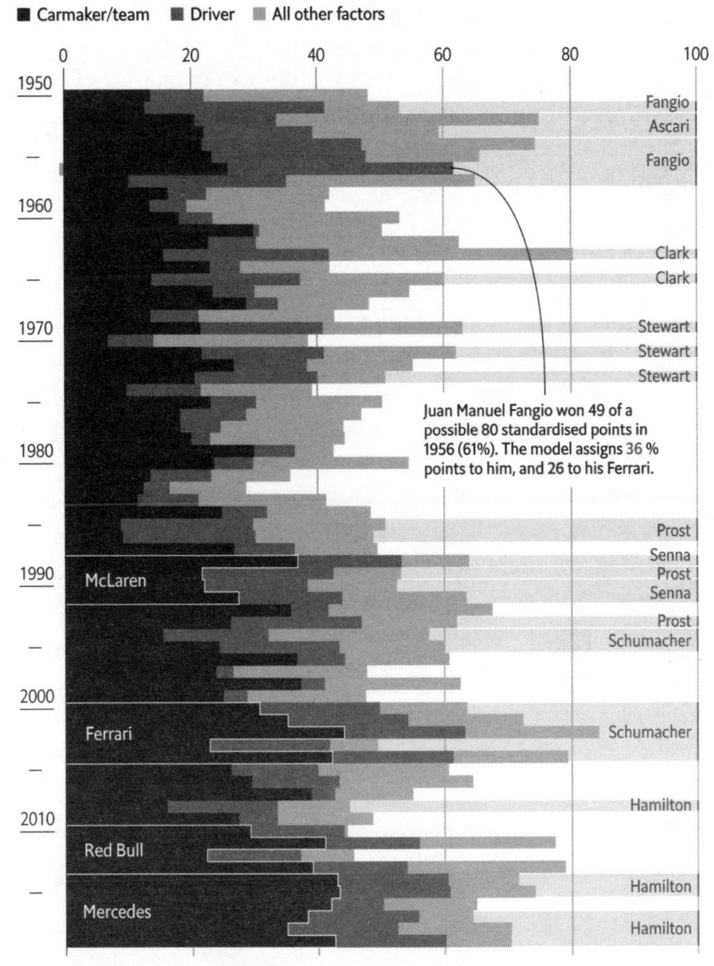

Sources: Ergast.com; f1-facts.com; "Formula for success: multilevel modelling of Formula One driver and constructor performance, 1950–2014", A. Bell et al., *Journal of Quantitative Analysis in Sports*, 2016; *The Economist*

*Ten points for first place, six for second, four for third etc.

Schumacher. He won 1.9 more points per race than an average driver would have done in the same events and cars, edging out Mr Hamilton's score of 1.8. Limited to their five best consecutive years, the gap widens, to 2.7 points per race for Mr Schumacher and 2.0 for Mr Hamilton. This difference stems mostly from the impact of their cars. Both stars raced in the finest vehicles of their day. But 20 years ago, cars from Williams and McLaren were nearly as strong as Ferrari's. In contrast, Mercedes now towers over its rivals, enabling Mr Hamilton and Valtteri Bottas, his teammate, to coast past lesser cars. Before joining Mercedes, Mr Bottas had never won a F1 race. He now has nine victories.

Yet on a per-race basis, the greats of yesteryear beat both modern stars. Three of the model's top four drivers had stopped racing by 1973; the leader, the Argentine Juan Manuel Fangio, won five titles in the 1950s. These pioneers had short careers. Fangio started just 51 races, to Mr Schumacher's 306. However, the model is impressed by them, because the impact of cars relative to drivers has grown over time. On average, it assigns drivers in the 1950s 58% of their teams' points; today, that share is 19%. Fangio, who was a mechanic by training and won titles using cars from four different firms, was known as "the master". The masters of modern F1 are engineers who sit behind laptops, not steering wheels.

Why young chess stars always usurp the old

From the very first episode of *The Queen's Gambit*, a hit Netflix miniseries about chess in the 1960s, it is clear what a precocious talent Beth Harmon is. Before her tenth birthday, she has learned to beat the janitor at the orphanage in Kentucky where she lives. Soon she takes on an entire college chess club in simultaneous matches, winning each one easily. By the time she reaches her troubled teenage years, she is vanquishing all comers, including stalwarts who are considerably older. After Beth wins a gruelling two-day match against one silver-haired champion, he gracefully concedes: "You are a marvel, my dear. I may have just played the best chess player of my life."

The seven-episode drama has received universal acclaim from critics: of the 58 reviews gathered by Rotten Tomatoes, an entertainment website, every one was positive. But it has also been praised by chess aficionados for its accuracy (doubtless helped by having Garry Kasparov, a former world champion, as a consultant). And an analysis by three economists confirms that the series' portrayal of a young upstart vanquishing her elders is exactly what happens in real chess, decade after decade.

The study, published in October 2020 in *Proceedings of the National Academy of Sciences*, analysed 24,000 matches involving world champions between 1890 and 2014. To assess the performances of the champions and their opponents, the academics compared their 1.6m moves against Stockfish 8, a chess-playing program that computes the best possible move for a given configuration of pieces on the board. The players were scored according to how often they picked Stockfish 8's optimal moves. (The researchers also estimated how each move affected a player's chance of winning and how often they made catastrophic mistakes.)

These results produced two clear conclusions. First, players tend to reach their peak early in their careers, with little improvement after their 30s. (There are even signs of a decline after 50.) Second, each generation comes closer than the last to Stockfish 8's

Generation gap

Performance of world chess champions and their opponents

Share of moves that are optimal for a given configuration, %

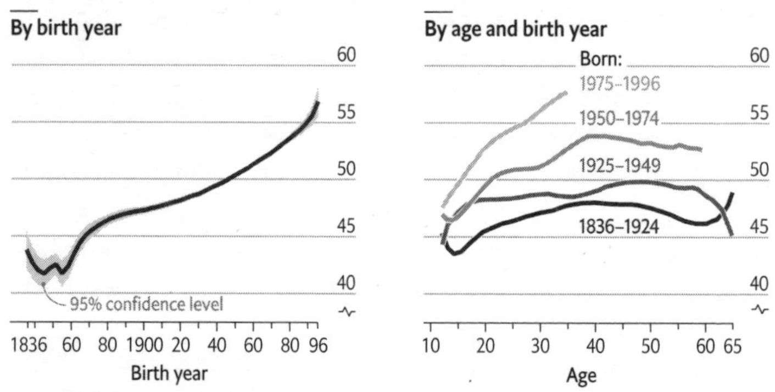

Source: "Life cycle patterns of cognitive performance over the long run", A. Strittmatter, U. Sunde and D. Zegners, *PNAS*, 2020

benchmark of optimal play. Professional players born in the 1950s had already reached a higher average level of performance by the age of 25 than those born in the 1920s ever did.

The authors reckon that the early-peak effect can be explained by the fact that the human brain's problem-solving ability (or "fluid intelligence") reaches its high point at around the age of 20. As for the long upward trend in performance through the decades, the authors suggest that more rigorous training is probably the cause. Indeed, modern chess masters can study the machines that now tend to beat them. If Beth were playing a methodical young champion today, she might surprise them with unorthodox play – but she would make enough mistakes to lose most of her games.

Media matters: arts and culture

Why dead rappers now get a bigger sales boost

Hip-hop fans are all too familiar with the success that can come after an artist's untimely death. Tupac Shakur and Notorious B.I.G., two American rappers who were murdered in 1996 and 1997 respectively, have sold more music in death than in life. Other well-known rappers to notch up hits after their deaths include Eazy-E (who died in 1995), Big L (1999) and J Dilla (2006). The past few years have seen a flurry of such posthumous hits. Juice WRLD, a rapper who died in December 2019, reached the top of America's Billboard 200 charts for the second time the following year with his third album, *Legends Never Die*. By one reckoning, it was the most successful posthumous release in two decades.

An analysis by *The Economist* suggests that, in the world of hip-hop at least, the sales boost generated by posthumous albums may be growing. We looked at releases by hip-hop artists Lil Peep, XXXTentacion, Mac Miller, Pop Smoke and Juice WRLD. To measure the commercial success of a release, we used the album-equivalent unit (AEU), a measure developed by Billboard and Nielsen SoundScan, a research firm, which treats 1,500 song streams or ten song downloads as equivalent to an album sale. To avoid comparing albums released before and after the adoption of the AEU in 2014, we restricted our analysis to those released in the years since.

All five posthumous albums in our sample performed better in their first week than previous works by the same artists. Pop Smoke and Juice WRLD's posthumous albums, both released in July 2020, amassed roughly four and seven times more AEUs, respectively, than the average releases during their lives. The posthumous works of Lil Peep and Mac Miller also recorded huge jumps in first-week sales.

Critics of posthumous releases – including fans, music critics and artists alike – say they are a corporate cash-grab and a blight on a dead artist's career. They allege that record labels compiling posthumous releases cram them with filler tracks and unfinished songs, in part to boost sales figures (the AEU system favours

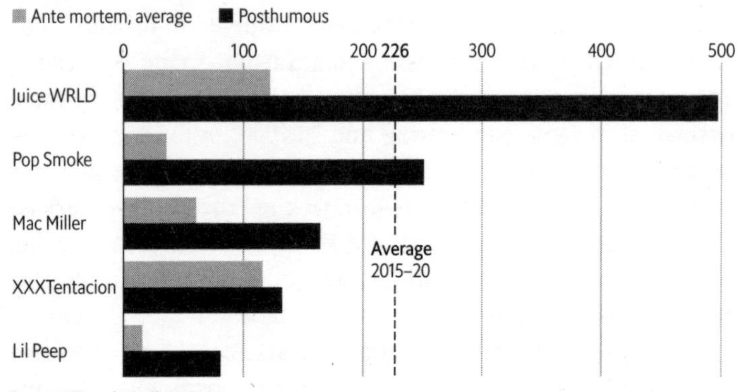

releases with more songs). There may be some truth to that. Since 2018, according to our figures, there has been a positive correlation between the number of tracks on a chart-topping album and its AEUs during the first week. But the relationship is very weak.

So why do posthumous albums often outperform their ante-mortem predecessors? The answer may be humdrum. Fans rally to their beloved artists. The day after David Bowie's death in 2016, the rock star's streams on Spotify surged to 2,700% of their typical levels. Living musicians can benefit, too. Artists who re-emerge after a hiatus, as Tool, a Los Angeles rock band, did in 2019 after 13 years without releasing an album, have seen their discographies appear on the Billboard charts. But some artists, unfortunately, achieve their greatest success only after the mic cable has been severed for good.

Why musicians are increasingly collaborating with their fans

On April 6th 2020 Charli XCX, a British pop singer, shared her intention to create a "DIY Quarantine album". "I'm going to be setting up an email so fans can send me beats," she announced in a video posted on Twitter, "I'll also be reaching out to collaborators online, and will keep the entire process super open." In the weeks leading up to the record's release on May 15th, she shared demos with her fans, asked them for feedback on lyrics and requested that they send in video clips of themselves so they could appear, uncredited, in a music video for one of the album's singles. The result, entitled *How I'm Feeling Now*, became one of the most talked-about albums of lockdown and reached number 33 in the British charts – an impressive feat, considering the avant-garde and chart-averse nature of the album.

Charli XCX encouraged her fans to regard themselves not as mere consumers of her work, but as her creative partners. Other musicians, sensing that their admirers needed distraction and entertainment in lockdown, had similar ideas. In August 2020 Snow Patrol, a Northern Irish band, released an album in collaboration with The Saturday Songwriters – a catch-all term for the fans who helped craft the songs via Instagram Live. Gary Lightbody, the band's lead singer, asked participants to suggest chords; he crafted melodies and users voted for their favourites. Mr Lightbody then solicited ideas for lyrics. "Some people were making jokes," he said, "but 95% of them were really heartfelt people speaking from where they were in that time and that place." The album's artwork was also created through an online contest, and all proceeds from the release went to a British food charity.

Social media played a crucial role in facilitating such dialogues and allowing creativity to flourish. Some musicians are making use of the "duet" function on TikTok, a video-sharing app, which allows users to riff on others' work. (If a user posts a video, another person can respond to it, and the results appear side-by-side on the screen.)

Charlie Puth, an American singer-songwriter and producer, began laying down beats on TikTok in February 2020 and asked users to add musical layers of their own. What resulted were fully fledged songs, which Mr Puth did not release but left on TikTok for people to enjoy. The hashtag #writethelyrics has been viewed more than 3.8bn times to date.

Some musicians sought fans' involvement in other ways. In April 2020 Grimes, a Canadian singer-songwriter, asked people to make a music video for her single "You'll Miss Me When I'm Not Around". "Because we're all in lockdown, we thought if people are bored and wanna learn new things, we could release the raw components of a music video for anyone who wants to try making stuff using our footage," she said. The finished products were posted on YouTube, Twitter and Instagram under the hashtag #grimesartkit, and some were shared by Grimes herself. In May Perfume Genius, an American singer, invited fans to submit ideas for the video for his single "Without You", with the three chosen directors receiving $1,000 to make their vision a reality.

"When I was growing up, I wanted to connect with the artist. I think the tools have just caught up with what fans have always wanted," says Eric Fritschi, a global marketing strategist who has worked on promotional campaigns for the likes of Diplo and Major Lazer. As well as making fans feel closer to their favourite singers, these schemes have benefits for record labels. "With the old model, you'd make all the content and storytelling, push it out to people and hope that they liked it, and there was very little feedback from that," he says. "But now we know, even before it goes out whether fans will like it or not, whether they're anticipating or not." Mr Fritschi estimates that a marketing team now makes about a tenth of an album's promotional material and "the fans are contributing all the rest" – either by posting on Twitter, dancing to the songs on TikTok or using the music to accompany an Instagram story.

Fans, no matter how many hours they put into these projects, do not have any formal claim to the remixed content. Copyright laws "prevent the [musicians'] work from being exploited in a way which they do not agree with, or in a way which means they are no longer appropriately remunerated for their work," says Joshua Schuermann, an associate at Briffa Legal, a firm specialising in intellectual property law. In 2014 Erasure, an English synth-pop duo, launched a competition in which fans could put their own spin on the band's latest single. The winning entry, "Sacred (Fiben Remix)", appeared on a record a year later, but its maker was not entitled to monetary compensation. Superfans may relish the opportunity to engage with their favourite tracks – but it is often for the musician's benefit, not theirs.

How the cola wars became a cultural phenomenon

"Rock and roller cola wars, I can't take it any more!" cried Billy Joel in his chart-topping song from 1989, "We Didn't Start the Fire". He had had enough of the intense marketing battle between America's fizzy-drinks behemoths. As the underdog, PepsiCo had stunned its bigger rival, Coca-Cola, by signing Michael Jackson, the era's biggest musical star, to promote its brand in a record-setting $5m deal.

How did the cola wars became such a cultural phenomenon? Credit for that goes to Donald Kendall, PepsiCo's legendary former boss, who died on September 19th 2020 aged 99. A gifted salesman, he rose quickly through the ranks from his start on the bottling line to become the firm's most senior sales and marketing executive at the tender age of 35. Seven years later he was named chief executive. In 1974 he injected a dose of fizzy capitalism into the Soviet Union, which allowed Pepsi to become the first Western product to be legally sold behind the iron curtain. By the time Mr Kendall stepped down as boss in 1986, PepsiCo's sales had shot up nearly 40-fold, to $7.6bn. His legacy continues to shape the industry.

Mr Kendall offered a mix of strategic vision, principled leadership and marketing flair. Two years after taking charge he acquired Frito-Lay, a leading purveyor of snacks, giving PepsiCo an advantage from diversification that persists to this day. PepsiCo's revenues in 2019 of $67bn dwarfed Coca-Cola's $37bn in sales. Decades before Black Lives Matter, Mr Kendall named African-Americans to top jobs, making PepsiCo the first big American firm to do so – staring down racists including the Ku Klux Klan, which organised a boycott.

But his masterstroke was the all-out marketing blitz against Coca-Cola, long the global market leader in non-alcoholic beverages. The two firms had competed for decades, but they mostly fought low-grade battles. Mr Kendall changed that, by forcing both companies into an advertising arms race. In 1975 Coca-Cola spent around $25m on advertising and PepsiCo some $18m. By 1985 those figures had shot up to $72m and $57m respectively. In 1995 Pepsi outspent Coke by $112m to $82m.

This was a risky gambit for both cola rivals. But it paid off in two ways. First, it helped fizzy drinks win a greater "share of throat" (a term coined by Roberto Goizueta, a former boss of Coca-Cola, who died in 1997). They went from 12.4% of American beverage consumption in 1970 to 22.4% in 1985. And although Coca-Cola maintained its lead in that period, with over a third of the market, PepsiCo's share rose from 20% to a peak of over 30% in the 1990s. In 2019 carbonated-drinks sales totalled $77bn in America, and more than $312bn globally. Coca-Cola and PepsiCo remain dominant.

The second way that the cola wars benefited both companies was by turning them into "the world's best marketers", observes Kaumil Gajrawala of Credit Suisse, a bank. Today a decades-long obsession with volume growth has been replaced by a focus on revenues and profits. PepsiCo in particular has relinquished some of the soft-drinks market, where its share has fallen back down to a quarter. But its marketing magic continues to sparkle, even if it is now deployed to sell less sugary alternatives such as bottled water, coffee and energy drinks to health-conscious consumers. In many industries a cosy duopoly retards innovation and harms consumers. The happy outcome of the cola wars has been the exact opposite. As Mr Kendall himself observed, "If there wasn't a Coca-Cola, we would have had to invent one, and they would have had to invent Pepsi."

How Hollywood is losing ground in China

Western audiences will be familiar with at least some of entries on the list of China's favourite films of 2019. It includes the superhero flicks *Avengers: Endgame* and *Spider-Man: Far from Home*, along with the latest instalment of the *Fast & Furious* franchise. Other titles may be less recognisable. *Ne Zha*, an animated fantasy-adventure, made more than $700m at the box office. *The Wandering Earth*, a sci-fi thriller, brought in $690m. Of the ten highest-grossing films in China in 2019, only three came from Hollywood. The rest were made locally.

The bounty of home-grown blockbusters in Chinese cinemas is relatively new. For years, moviegoers preferred Hollywood features. In 2007, 14 of the 25 highest-grossing films in China were made in America. Yet as China's box office has grown – receipts reached $9.2bn in 2019, up from just $800m a decade ago – mainland studios have captured a bigger share. In 2019, 17 of the top 25 films in China were Chinese. Only eight were American.

Tinseltown is not taking this lying down. With China on track

Bamboo screen
China

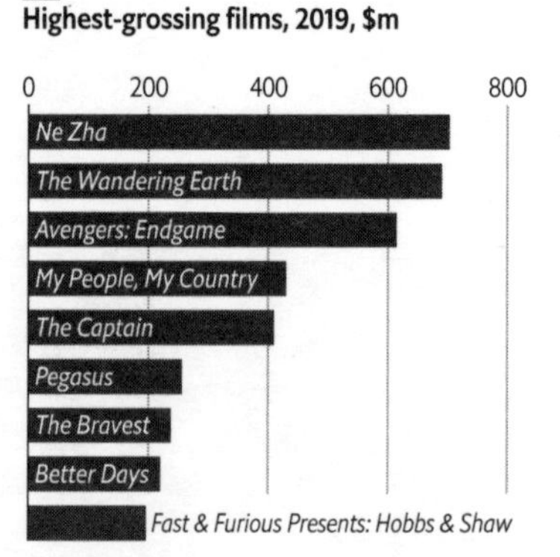

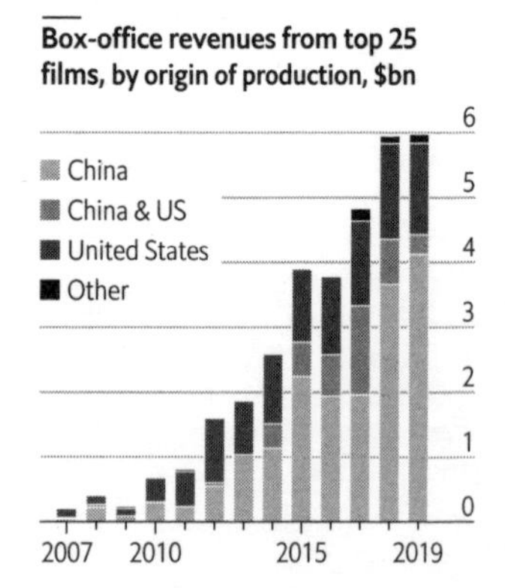

Sources: Box Office Mojo by IMDbPro; ENDATA

to become the biggest film market in the world, Hollywood studios are tailoring their films to suit Chinese tastes. Disney's live-action remake of *Mulan*, released in 2020, was one example. In an early version of the script, the title character shared a kiss with her love interest, but Disney cut the scene after Chinese executives deemed it inappropriate for local audiences. Such censorship is likely to continue. Western audiences may prefer American films to Chinese ones, for now. But soon this distinction may have little meaning.

What accounts for the strange appeal of snow globes?

The snow globe is a peculiar object, something between toy and decoration. Wintry scenes suspended in liquid may be whimsical but they are not just for children. Among their most famous collectors was Walter Benjamin, a German philosopher. The small "glass balls containing a landscape upon which snow fell when shook," his friend Theodor Adorno observed, "were among his favourite objects". In Europe in the 1930s the snow globe had not yet become the model of kitsch it is today. But Benjamin was fascinated by the quasi-magical appearance of these increasingly popular commodities, whose blustering spectacles might represent the imagination in flux or the maelstrom of modern artifice. Such ambiguities have imbued the snow globe with a rich symbolic potential, which explains why they have so often enjoyed starring roles in visual culture.

Today they are often maligned as sentimental bric-a-brac. But a quick glance at the snow globe's history reveals close ties with the rise of modernity. The earliest models were showcased at the Paris Universal Exposition in 1878; they reappeared again at the event in 1889 that marked the unveiling of the Eiffel Tower. This crystallised the snow globe's associations with cutting-edge design and, crucially, the glamour of travel. The novelty soon made its way to America. Plastic replaced the glass domes, allowing the keepsakes to be mass produced and transported internationally; by the early 1940s they were everywhere.

So ubiquitous was the snow globe that it began to infiltrate Hollywood, most famously at the beginning of Orson Welles's *Citizen Kane* (1941). On his deathbed, the reclusive publishing tycoon clutches a dome featuring a diorama of a snow-topped log cabin. With his final breath, it falls to the floor and smashes, severing his last connection to the cherished memories of his youth. As bridges to the past, snow globes have been used to explore states of reverie and yearning in film almost to the point of cliché.

The year before the release of *Citizen Kane*, Ginger Rogers starred in *Kitty Foyle*, in which her character's encounters with a snow globe trigger flashbacks to her childhood. The film cemented the item's status as a must-have possession in American homes, causing a tripling of sales in the United States.

For all their festive appeal, snow globes rarely appear in film without a tinge of melancholy. Many directors have probed the dark undertones of an idealised world sealed off from the outside. In *The Lovely Bones* (2009), the snow globe is imagined as a dreamlike purgatory where the soul of murdered 14-year-old Susie Salmon is trapped. It was not the first time film-makers had linked the memento to murder. Snow globes were used as promotional items for the Coen brothers' black comedy *Fargo* (1996). Those who bought the special edition video set received a globe depicting policewoman Marge Gunderson crouching over a dead body.

Visual artists have also found value in subverting the snow globe's anodyne cheeriness. The decoration caught the eye of Surrealists; in *Boule sans neige* (1927), the American artist Man Ray replaced a snowy idyll with a single eye, reversing the gaze between object and viewer. LigoranoReese (the moniker of a pair of artists) enclose curse words in their snow globes, garbling the souvenir's wholesome message. Another duo, Martin & Muñoz, bring their own unsettling spin to the form by crafting eerie frigid landscapes populated by dead trees and tiny humans in absurdist scenarios: carrying gravestones, trapped under houses or about to be eaten by giant spiders.

Yet the snow globe's fairy-tale connotations persist in mainstream culture, particularly in pop music. In the music video for "Lover" (2019), Taylor Swift plays out romantic fantasies in a house encased in a snow globe. The final shot reveals a festive family scene in which a child marvels at the newly gifted dome. The ornament's reputation for instant gratification and neatly packaged charm is clear, creating a perfect snowstorm of mass-market pop and Christmas consumerism. As a self-contained world, one suspended between stillness and movement, life and lifelessness, the snow globe has particular appeal in times of upheaval. In periods of great turmoil, as Benjamin, film-makers and artists have found, it seems to represent the vicissitudes of life.

How does Christmas television vary around the world?

Even in the age of streaming, the announcement of the Christmas schedule on terrestrial television is met with excitement. It is a time of year when friends and families gather round the set, glass of mulled wine or fizz in hand, and watch something together: in 2019 a special festive episode of *Gavin and Stacey*, a sitcom, became the most-watched scripted show in Britain in a decade. Broadcasters usually put together a mix of new fare, highlights shows and old classics to get viewers in the spirit. It is inevitable that, in Britain and America, *It's a Wonderful Life* (1946), *White Christmas* (1954), *Elf* (2003) or *Love Actually* (2003) will feature in the line-up in the countdown to Christmas.

But what about elsewhere in the world? Period dramas are a must on the schedules of Swedish and German channels. Swedes' favourite at Yuletide is *Fanny and Alexander* (1982), directed by Ingmar Bergman. It begins with one of the most evocative Christmas celebrations ever filmed as the titular siblings enjoy a meal with their mother, Emilie, and father, Oscar. The merriment becomes the stuff of nostalgia after Oscar's sudden death and, though the children suffer at the hands of a new puritanical stepfather, they prevail in the end. Germans, meanwhile, watch *Sissi* (1955) in which Romy Schneider played the Austrian princess who captured the heart of the emperor Franz Joseph. The film is a fairy-tale rooted in real life, involving the same royal pageantry that British viewers may recognise from their own ritual watching of the queen's Christmas broadcast.

For Italians, Advent would be incomplete without the farces known as *cinepanettone* (named after the sweet bread traditionally eaten at Christmas time). Often starring an ensemble of Italian stars, *cinepanettone* are cheerful, crude and absurd, focusing on the sex lives of their characters while on their Christmas break. The tradition began with *Vacanze di Natale* ("Christmas Holidays") in 1983, which followed the romantic entanglements of an upper-crust family from Milan and a humbler family from Rome. It was a hit at the box office, spawning sequels in 1990, 1991, 1995 and 2000, as well as similar stories set in far-flung places. In 2020 Neri Parenti, a stalwart of *cinepanettone*, released *In vacanza su Marte* ("On Vacation on Mars"). Critics loathe the genre – one compared the films to a filthy scrawl on the wall of a train-station toilet – but it endures all the same.

France's festive film of choice is also a comedy, though one with much darker undertones. *Le Père Noël est une ordure* ("Santa Claus is a Stinker"), a play which was adapted into a film in 1982, follows Pierre and Thérèse, two volunteers at a suicide-prevention helpline, on their tumultuous Christmas Eve shift. The film appeals to Gallic irreverence and delights in blowing a hearty secular raspberry at the sentimentality of Christmas. So popular is the film that Nora Ephron and Delia Ephron wrote an English-language remake in 1994, but it was panned by critics and made a substantial loss at the box office.

A firm favourite in Russia and the former Soviet republics is an adaptation of a Hans Christian Andersen's fairy-tale, *The Snow Queen* (1957), directed by Lev Atamanov and Nikolay Fyodorov and featuring glorious hand-drawn animations. More surprising is the popularity of *The Irony of Fate* (1976), a three-hour-long comedy which follows the plight of Zhenya, a young Muscovite doctor. After a drunken party in a bathhouse with his friends, Zhenya finds himself in Leningrad: the joke is that, given the uniformity of Russian architecture, the protagonist doesn't notice. Though the film is subversive – implying that individuality has been subsumed by bland anonymity – it slipped past the censors, at least until Mikhail Gorbachev took charge. The Soviet leader saw the film as encouraging alcoholism and had it banned, a prohibition that was as doomed to failure (videotapes were already available) as his abstinence campaign. In Poland *Home Alone* (1990) gets an annual airing; in Estonia not only is *Die Hard* (1988) considered a bona fide Christmas film, but the whole series is shown on consecutive nights, starting on Christmas Eve.

Some of these films are critically acclaimed, but at Christmas the ritual is more important than the substance (the nostalgia and sense of continuity from one year to the next must be the only reasons people still bother with *Love Actually*). Just as the purpose of Christmas pudding is not nutrition, and Christmas crackers are not pulled to provide stylish headgear, so festive movies are not chosen for their quality. They are for making and reliving shared memories – and slowly digesting all that turkey.

What is the world's most frequently aired television programme?

Dinner for One is watched all over the world on December 31st. The 18-minute British comedy sketch, recorded in 1963, holds the Guinness World Record for the most frequently aired television programme. It is a particularly crucial part of Germany's festive programming, where it has been broadcast on Das Erste, a public television channel, since 1972. Around half the population still tunes in on New Year's Eve to watch it; some Germans recreate the meal served in the skit. In 2018 the German Post Office unveiled a set of commemorative stamps featuring its characters. One of its lines – "The same procedure as every year" – has become something of a catchphrase in the country. So enduring is *Dinner for One* that in 2016 Netflix released a parody of it.

Filmed in grainy black-and-white, the routine involves Miss Sophie (May Warden), who is celebrating her 90th birthday with James, her butler (Freddie Frinton). She has outlived her four closest friends and so insists that James impersonate them all in turn. They are an eclectic bunch. First is Sir Toby, a raspy-voiced northerner. Next up is Admiral von Schneider, who raises his toasts with a loud "Skol!" Then comes the turn of Mr Pomeroy, who speaks in an alarmingly high-pitched falsetto. The last guest is Mr Winterbottom, Miss Sophie's "very dear friend", who is a booze-swilling lecher. With every course served and glass raised in Miss Sophie's honour, James gradually becomes so drunk that he can hardly stand (in an extra irony, it so happens that Frinton himself was teetotal). Drinks are spilled, plates go flying and, in a recurring gag, James keeps tripping over the head of an enormous tiger rug. Miss Sophie, seemingly oblivious to the chaos around her, ploughs on and has a thoroughly enjoyable evening. It soon becomes clear that this performance happens every year.

Part of the sketch's appeal is that it features someone getting stinking drunk. This may not be an original scenario in comedy, yet there is something undeniably charming about the interplay between a kindly, if bossy, elderly woman and her blindly loyal manservant. The stiff social setting is brilliantly contrasted with the cast of eccentric characters and the surrealism of the whole imaginary set-up. There are some obvious reasons why it is popular even among those who do not speak English. The comedy is mostly physical, and easily breaks through the language barrier; the innuendo-laden finale is not hard to parse, either.

What is curious is how little-known the sketch is in its country of origin. Though it was first performed in Britain as early as 1934, today hardly any Britons have heard of or seen it. Its first airing on national British television was in 2018, more than half a century after it was filmed, and even then on Sky Arts, a pay-TV channel. (It is now also freely available on YouTube.) Some critics have argued that Britons shun it due to discomfort with the sketch's depiction of class, though this seems unlikely given the sending-up of snobbery in some of the country's best-loved comedies such as *Dad's Army* and *Monty Python*. Other commentators think it is the references to empire, such as that tiger-skin rug, that sit uncomfortably with Brits.

If it is largely ignored by Britons, *Dinner for One* has a loyal and enthusiastic audience elsewhere – as well as Germany, it is shown during the festive period in Estonia, Australia, Norway and Denmark. At the turn of the year people gravitate towards old television shows and films as a form of comfort. *Dinner for One* links people to the past and, perhaps, brings back fond memories of loved ones of their own who are no longer here.

Contributors

THE EDITOR WISHES TO THANK the authors and data journalists who created the explainers and accompanying graphics on which this book is based:

Nour Abraham, Olivia Acland, Miranda Aldersley, Helen Atkinson, Georgia Banjo, Dylan Barry, Hollie Berman, John Bleasedale, Adrian Blomfield, Aryn Braun, Sophia Caldera, Bruce Clark, Philip Coggan, Simon Cox, Rachel Dobbs, Kristina Foster, Bo Franklin, Lane Greene, Amy Hawkins, Clara Hernanz, Mike Jakeman, Shashank Joshi, Matt Kaplan, Andrew Knox, Léa Legraien, Sarah Leo, Matt Lerner, John McDermott, Matt McLean, Clare McQue, Roger McShane, Emma Madden, Jessie Mathewson, Adam Meara, Sacha Nauta, Lloyd Parker, Rosamund Pearce, Sophie Pedder, Lizzy Peet, Nick Pelham, Clara de Pretis, Simon Rabinovitch, Zamira Rahim, Mike Reid, Adam Roberts, Kinley Salmon, Guy Scriven, Alex Selby-Boothroyd, Sondre Ulvund Solstad, Benjamin Sutherland, Boyd Tonkin, Kaitlin Tosh, James Tozer, Vijay Vaitheeswaran, Matthew Valencia, Hanna Vioque, Maria Wilczek, Jonny Williams, Olivia Williams, Sue-Lin Wong, Wade Zhou and Dominic Ziegler.

Index

A
Abdel Aziz, Mohamed Ould 93
Abe Shinzo 159, 160
Abiy Ahmed 62, 63
Abu Dhabi, mosques 83
academic journals 204–5
Academy Awards 183
adenovirus vaccines 124
Adorno, Theodor 236
Africa
  covid-19 fatality rate 121
  football and civil war 5–6
  gender life-expectancy gap 111
  population predictions 99
  visa costs 152–3
  see also individual countries
African Continental Free Trade Area (AfCFTA) 145–6
African Union (AU) 145
African-Americans
  at PepsiCo 232
  in DC 59–60
Akyol, Mustafa 81–2
al-Nahyans 83
Alabama, religion 86
Albania, freedom 81
album-equivalent units 227–8
alcohol
  Dinner for One 242–3
  and life-expectancy 111
  South Africa 33
Alexa 161
Algeria
  freedom 82
  mosques 83
  and Western Sahara 42
Allam, Shawki 79, 80
Amazon 196, 197
Alexa 161
Amazon Prime Day 144
America *see* United States
amphetamine, and coronavirus pandemic 13
Andersen, Hans Christian 240
"Animal Crossing: New Horizons" 136–7
ant-egg soup 39–40
Antifa 64
Apple 149, 161
ArcelorMittel 191
Arctic
  maritime boundaries 68–9
  sea ice 189–90
Armenia 8

armies 49–51
  and democracy 71
The Art of the Deal (Trump) 36
The Art of War (Sun Tzu) 36
AstraZeneca/Oxford vaccine 124
Atamanov, Lev 240
athletics 218–19
Atlanta, cost of living 45
ATM cash-out 19–20
Australia
  cannabis use 12
  Dinner for One 243
  forest fires 197
  population predictions 100
  women's employment 107
Austria, women's employment 107
Avengers: Endgame 234
AVEs 54–5
Azerbaijan, women's rights 81
Azhari, Osama al- 79

B

Bach, Johann Sebastian 211
Bahamas, sand dollar 149
Bahrain, freedom 82
Baltic states
  electric grids 37–8
  see also Estonia; Latvia; Lithuania
Balzac, Honoré de 31
Bangladesh
  banks 19, 20
  blasphemy laws 93
  democracy 71
  freedom 81, 82
Bank of Mum and Dad 23–5
banks
  central-bank digital currencies 149–51
  cybercrime 19–20
bark beetles 197
BeagleBoyz 19–20
Belarus
  and Baltic states 37, 38
  democracy 71
  digital censorship 76
  gender life-expectancy gap 111
Belgium, women's employment 107
Bell, Andrew 220
Benjamin, Walter 236, 238
benzodiazepines, and coronavirus pandemic 13
Bergman, Ingmar 239
Better Days 234
Bhutan, gender life-expectancy gap 112
Bible, dogs 80
Biden, Joe
  Catholicism 85
  and DC statehood 60
  immigration policy 115
  speaking style 166–7
  and white supremacy 64
Big L 227
Big Mac index 16–18
biological evolution 168–70
Björk, Bo-Christer 205
Black Friday 143–4
black holes 200
blasphemy laws 93–4
Blavatskyy, Pavlo 7–8
Blinken, Anthony 180
blursday 143

body-mass index (BMI), and corruption 7–9
Bolsonaro, Jair 46, 47
books, self-help books 35–6
Bosnia
  freedom 81, 82
  genocide 181
  women's rights 81
Boston, cost of living 45
Bottas, Valtteri 222
bottle deposit refund schemes 187–8
Boule sans neige 237
Bouteflika, Abdelaziz 83
Bowie, David 228
The Bravest 234
Brazeau, Martin 202–3
Brazil
  corruption 46–8
  economy 17
  education 15
Brewer, Rosalind 107
Brexit, fishing rights 69
bribery 154–6
bristlecone pine 195
Britain
  army 49, 50
  Bank of Mum and Dad 23
  cannabis use 12
  Christmas television 239
  Dinner for One 242
  economy 17
  fertility rate 115
  fishing rights 69
  nurses 126–8
  period poverty 102–3
  royal family 52–3
  stamps 211, 212
  video gaming 136–7
  waste 187
  women's employment 107, 108
  see also England
Broadie, Mark 217
Brueghel, Jan 211
BTS 165
Buchan, James 127
Buckingham Palace 52, 53
Buddhism
  American politicians 87
  in China 90
Bulgaria, population predictions 100
Burkina Faso, freedom 81
Burmese language, Wikipedia 177
Byers, Michael 68–9

C
Cabells 204
California
  giant sequoias 197
  population 114
  road deaths 119–20
Cambridge, Duke and Duchess of 53
Canada
  fertility rate 115
  gender pay gap 110
  maritime boundaries 69
  women's employment 107
cancer deaths 134
cannabis, and coronavirus pandemic 12, 13
Capek, Karel 161–3
The Captain 234
cars
  number plates 31–2
  road deaths 119–20

cash-out schemes 19–20
Catholicism
  American politicians 85, 86
  in China 90, 91
cats 80
Cebuano, Wikipedia 178, 179
Central African Republic (CAR), stamps 211, 212
central-bank digital currencies (CBDCs) 149–51
cerebrovascular mortality 133, 134, 135
Chad, freedom 82
Chamie, Joseph 116
Charles, Prince 53
Charli XCX 229
Charpentier, Emmanuelle 200
Chen Duxiu 160
Cheptegei, Joshua 218
chess 223–4
Cheung, Yan-Leung 154
Chile, maritime boundaries 68
China
  Christianity 90–92
  covid-19 fatality rate 121, 122
  democracy 71
  digital yuan 149
  economy 16–18
  films 234–5
  gender life-expectancy gap 112
  geographical economic inequalities 141–2
  hair 10, 11
  heatwaves 129, 130
  maritime boundaries 68
  population predictions 100–101
  self-help books 35–6
  Singles' Day 144
  Sinovac and Sinopharm vaccines 123
  and Taiwan 73–4
  Uyghurs 91, 180, 181
Chinese language
  reform 160
  Wikipedia 178
Chini, Louise Parsons 196
chondrichthyans 202
Christianity
  in America 85, 86
  in China 90–92
  and dogs 80
Christmas
  snow globes 236–8
  television 239–41
cinepanettone 239–40
Citicorp 107
cities, cost of living 44–5
Citizen Kane 236
civil list 52
civil war
  Ethiopia 62–3
  and football 5–6
climate change, and Arctic sea ice 189–90
climbers 193–4
Coca-Cola 232–3
cocaine, and coronavirus pandemic 12, 13
Coe, Sebastian 219
companies
  bribery 154–5
  emissions 191–2
Confucius 159
Confucius from the Heart (Yu Dan) 36
Congress, religious affiliation 85–7

Copenhagen, cost of living 44
corporate bribery 154–6
corruption
  Brazil 46–8
  and politicians' BMI 7–9
  South Africa 33–4
Corsica, number plates 31–2
cost of living 44–5
Costa, Paulo Roberto 46
covid-19 pandemic
  academic papers 204
  America 115
  Black Friday 143–4
  China 141, 142
  and cybercrime 19
  and drugs 12–13
  and football 209
  gender mortality gap 111–12, 113
  infection fatality rates 121–2
  and music 229–30
  nurse deaths 126
  remote working 27
  and road deaths 119–20
  Taiwan 73
  and turkey-meat trade 132
  Twitter posts 164, 165
  vaccines 123–5
  and video gaming 136–7
Crews, Christine 98
cricketers, on stamps 211
crime, digital bank-robbery 19–20
Croatia, education 15
Croijmans, Ilja 175–6
crown estate 52
cryptocurrencies 149
Crystal Palace 214
Cunha, Eduardo 47
Curtis, Jamie Lee 171
Cyber Monday 143, 144
cybercrime 19–20
Czech Republic, women's employment 107
Czechoslovakia, robots 161–3

D
Damascus, cost of living 45
Darwin, Charles 168
Daube, Mike 204
DC statehood 59–61
de la Fuente, Ángel 54
deaths
  on Everest 193, 194
  from covid-19 121–2
  heat-related 129
  on roads 119–20
Deaton, Angus 147
Debretsion Gebremichael 62
DeCar, Ron 97
DeChambreau, Bryson 216, 217
democracy
  and bribery 154, 156
  and internet access 75–6
  support for 70–72
Denmark
  Bank of Mum and Dad 23–5
  Dinner for One 243
  maritime boundaries 68, 69
  women's employment 107
Depetris-Chauvin, Emilio 5–6
deposit refund schemes 187–8
The Descent of Man (Darwin) 168
Diana, Princess 53
Die Hard 240
diet, Japan 133–5
digital currencies 149–51
Dinner for One 242–3

Disney 235
dogs 79–80
Doshay, Harris 91–2
Doudna, Jennifer 200
drinks
cola wars 232–3
see also alcohol
driving, road deaths 119–20
Drogba, Didier 5, 6
drugs, and coronavirus pandemic 12–13
Duckett, Thasunda Brown 107
dung beetles 198–9

E

East Germany
women's employment 104–6
see also Germany
Eazy-E 227
economy
Big Mac index 16–18
China 16–18, 141–2
ecstasy, and coronavirus pandemic 12, 13
Ecuador
maritime boundaries 68
visa costs 152
education
and population predictions 99
and wealth 14–15
Egypt
blasphemy laws 93
freedom 81, 82
Islam and dogs 79–80
mosques 83
Einstein, Albert 200
El Salvador, gender life-expectancy gap 111, 112
electric grids 37–8
Elf 239
Elizabeth, Queen 53
emissions, and green investment 191–2
employment
gender pay gap 109–110
German women 104–6
glass-ceiling index 107–8
England, football penalties 213–14
English language
irregardless 171–3
Wikipedia 177, 178
Ephron, Nora and Delia 240
Episcopalians 87
Erasure 231
Erignac, Claude 31
Eritrea 62, 63
and AfCFTA 145
Estonia
Christmas television 240, 243
electric grid 37
obesity and corruption 7, 8
eSwatini
gender life-expectancy gap 111, 112
non-tariff barriers 146
and Taiwan 74
Ethiopia, civil war 62–3
Europe
armies 49, 50–51
covid-19 fatality rate 121, 122
forests 196
see also individual countries
European Union (EU)
and Baltic states 37–8
digital euro 149
European Values Study 70, 71
Everest 193–4

evolution 168–70
ExxonMobil 191

F

F1 220–22
Fachin, Edson 46, 47
faeces 198–9
Fangio, Juan Manuel 221, 222
Fanny and Alexander 239
far-left terrorism 64, 65, 66
far-right terrorism 64–6
Fargo 237
Federal Republic of Germany (FRG) *see* West Germany
Federal Reserve 149, 150
Ferrari 220, 221, 222
fertility rates 99–100, 115
films
  Academy Awards 183
  China 234–5
  Christmas television 239–41
  snow globes in 236–7
Finland, women's employment 107
Floyd, George 64, 164, 165
fluid intelligence 224
food
  ant-egg soup 39–40
  and Japanese longevity 133–5
  turkey meat 131–2
football
  and civil wars 5–6
  penalties 213–14
  referees' bias 209–210
Foreign Corrupt Practices Act (FCPA) (US) 154, 155
forests 195–7
France
  army 50
  Christians 90
  Christmas television 240
  economy 17
  fertility rate 115
  laïcité 88–9
  number plates 31–2
  women's employment 107
Frankenstein (Shelley) 161
Franklin, Benjamin 131
Fraser, Jane 107
freedom
  blasphemy laws 93–4
  and Islam 81–2
French language, Wikipedia 178
Frey, William 114
Frinton, Freddie 242
Fritschi, Eric 230
Fukuzawa Yukichi 159–60
Fyodorov, Nikolay 240

G

Gajrawala, Kaumil 233
Gamaleya vaccine 124
Gaskins, Ben 86
Gavin and Stacey 239
gender pay gap 109–110
Geneva, cost of living 44
genocide 180–81
Georgia, obesity and corruption 7, 8
German Democratic Republic (GDR) *see* East Germany
Germany
  army 49, 50, 51
  Christians 90
  Christmas television 239
  deposit refund scheme 188
  Dinner for One 242–3
  economy 17

heat-related deaths 129
stamps 242
women's employment 104–6, 107, 108
see also East Germany; West Germany
Ghana, independence 145
giant sequoias 197
Gidey, Letesenbet 218
Giles, Elliot 218, 219
Ginet, Benoit 32
Giozueta, Roberto 233
glass-ceiling index 107–8
Goldstein, Joshua 115
golf swing 215–17
Google 177
Gorbachev, Mikhail 240
Greece, women's employment 107
green investment 191–2
Green, John 86–7
greenhouse-gas emissions, and green investment 191–2
Grimes 230
Guinea-Bissau, stamps 211
Guriev, Sergei 75–6

H

Hackett, James 50
hair, wigs 10–11
Hamilton, Alexander 59
Hamilton, Lewis 220, 221, 222
Hansen, Benjamin 109–110
happiness
and income 147–8
Twitter users 164–5
Harry, Prince 52, 53
Harvey, Ian 211
Hassan, Sinan 83
heatwaves
deaths 129–30
Siberia 189
Hebei 142
Hedonometer 164–5
Hemingway, Ernest 26
Hendriks-Kim, Eric 35–6
Higgs, Peter 200
High Wycombe 7
high-speed trains 54–5
Hindi, Wikipedia 178
Hinduism
American politicians 87
India 93
HIV/AIDS 111
Hollywood films 234–5
Home Alone 240
Hong Kong, cost of living 44
hostage negotiation 26–7
How I'm Feeling Now (Charli XCX) 229
Huffman, Jared 85, 87
Human Freedom Index 81–2
Hungary
democracy 71
women's employment 107
Hutton, Len 211

I

ice, Laptev Sea 189–90
Iceland, women's employment 107
Icelandic language 177
immigration
America 114, 115, 116
nurses 126–7
In vacanza su Marte 240
inactivated vaccines 123
income, and happiness 147–8

India
academic journals 204
blasphemy laws 93
covid-19 fatality rate 121, 122
economy 17
freedom 81
gender life-expectancy gap 112
hair 11
heatwaves 129, 130
nurses 126–7
population predictions 100–101
stamps 212
Indonesia
blasphemy laws 93
freedom 81, 82
heatwaves 130
maritime boundaries 67
Institute of Health Metrics and Evaluation (IHME) 99–101
International Football Association Board (IFAB) 213
internet, and faith in government 75–6
Iowa, speeding 119
Iran
blasphemy laws 93
dogs 80
freedom 81, 82
mosques 83
Iraq
democracy 71
freedom 81
Yazidis 181
Ireland, women's employment 107
The Irony of Fate 240
irregardless 171–3
Isaias Afwerki 63
Islam
American politicians 87
blasphemy laws 93
China 90, 91
on dogs 79–80
France 88–9
grand mosques 83–4
personal freedom 81–2
Israel, women's employment 107
Italy
army 49, 50
Christmas television 239–40
covid-19 fatality rate 121, 122
maritime boundaries 67
women's employment 107
It's a Wonderful Life 239
Ivory Coast, football and civil war 5–6

J

J Dilla 227
Jackson, Michael 232
Japan
covid-19 fatality rate 121, 122
economy 17
gender life-expectancy gap 112
heatwaves 130
longevity 133–5
names 159–60
women's employment 107, 108
Jefferson, Thomas 59, 85
Jehovah's Witnesses 93
Jermalavicius, Tomas 37–8
Joel, Billy 232
Johannesburg, restaurants 33
Johnson, Dustin 215, 216
Johnson, Spencer 35
Johnson & Johnson vaccine 124
Jones, William 168

Jordan, Michael 98
Jordan, freedom 82
journalism, Pulitzer prizes 182–3
Judaism
  American politicians 86
  on dogs 80
Juice WRLD 227–8
Julius Caesar (Shakespeare) 7

K
Kafka, Franz 162
Kahneman, Daniel 147
Kaliningrad 38
Kasparov, Garry 223
Kazakhstan
  freedom 82
  women's rights 81
Kearney, Melissa 115
Kendall, Donald 232, 233
Kenya, period poverty 103
Kerza, Edvinas 38
ketamine, and coronavirus pandemic 13
Khashoggi, Jamal 82
Killingsworth, Matthew 147–8
Kingston, Jeffrey 160
Kipchogen, Eliud 218
Kitty Foyle 237
Kuwait, blasphemy laws 93
Kyrgyzstan
  freedom 81
  women's rights 81

L
laïcité 88–9
Lancaster, Duchy of 53
language
  evolution 168–70
  Wikipedia 177–9
  wine vocabulary 174–6
  see also Chinese language; English language
Laos, ant-egg soup 39–40
Laperrière, André 131, 132
Laptev Sea 189–90
Latvia
  electric grid 37
  obesity and corruption 7, 8
Lava Jato 46–8
League of Nations, maritime boundaries 68
Lebanon, economy 17
Leicester City 213–14
Lee, Ronald 115
Legends Never Die (Juice WRLD) 227
Levine, Phillip 115
Libya, freedom 82
life expectancy 111–13
  America 115
  Japan 133
Lightbody, Gary 229
LigoranoReese 237
Lil Peep 227–8
Lin, Nay 11
Lincoln, Abraham 85
Linn, William 85
listening 26–7
Lithuania
  electric grid 37, 38
  gender life-expectancy gap 112
  obesity and corruption 7, 8
  recycling 187–8
A Little White Wedding Chapel 98
lockdown *see* covid-19 pandemic
longevity, Japan 133–5
Los Angeles, cost of living 44

Love Actually 239, 241
The Lovely Bones 237
"Lover" (Taylor Swift) 237
LSD, and coronavirus pandemic 13
Lu Xun 160
Lula da Silva, Luiz Inácio 46, 47
Lutz, Wolfgang 99

M
Maalouf, Amin 84
McDonald's, Big Mac index 16–18
McDowell, Nate 197
McLaren 221, 222
McNichols, Drew 109–110
Macron, Emmanuel 89
Madison, James 59
magic mushrooms, and coronavirus pandemic 13
Majid, Asifa 175–6
Malaysia, freedom 82
Mali, freedom 82
Man Ray 237
Manchester City 214
Manchester United 213, 214
Mangum, Kyle 115
maritime boundaries 67–9
marriage, Vegas weddings 97–8
Martin & Muñoz 237
Marukawa Tamayo 160
Masaryk, Tomas 162
Massachusetts, salary history ban 109
Mauritania, and Western Sahara 41–2
meat
  and cerebrovascular mortality 133–5
  turkey 131–2
Melnikov, Nikita 75–6
men, life expectancy 111–13
Mene, Wamkele 146
Mengistu Haile Mariam 62
menstruation 102–3
Mercedes 220, 221, 222
methamphetamine, and coronavirus pandemic 13
Mexico, democracy 70, 71
Miller, Mac 227–8
Minjinia turegensis 202
Mizumura Minae 160
mobile broadband 75–6
mobile phones, payment apps 149, 150
Moderna/NIAID vaccine 124
Morgan, John Tyler 60
Mori Arinori 159
Mormons 86
Moro, Sérgio 46, 47
Morocco
  education 15
  freedom 82
  and Western Sahara 41–3
mosques 83–4
motor sport 220–22
Mount Everest 193–4
Mourinho, José 213
Mozambique
  gender life-expectancy gap 111, 112
  non-tariff barriers 146
  stamps 211
Mulan 235
Mullender, Richard 26–7
music
  and cola wars 232
  fan involvement 229–31
  posthumous albums 227–8

and snow globes 237
Muslims *see* Islam
My People, My Country 234
Myanmar, hair 11

N
Nambu Yoichiro 200
names, Japan 159–60
Namibia, gender life-expectancy gap 111, 112
National Health Service, nursing vacancies 126
Ne Zha 234
Neri Parente 240
Netherlands, women's employment 107
New Year's Eve 242–3
New York
cost of living 44
population 114
New Zealand
democracy 70, 71
population predictions 100
and Taiwan 73
women's employment 107
Nicaragua
period poverty 103
and Taiwan 74
Niger
freedom 82
stamps 211
Nigeria
academic journals 204
blasphemy laws 93
freedom 82
gender life-expectancy gap 112
Twitter posts 164, 165
wigs 10–11
Nike 218
Nkrumah, Kwame 145
Nobel prizes 200–201
North Korea, cybercrime 19–20
North Pole 69
Norton, Eleanor Holmes 59, 60
Norway
Dinner for One 243
maritime boundaries 68
navy 50
women's employment 107
Notorious B.I.G. 227
Novavax vaccine 123
number plates 31–2
nurses 126–8

O
Obama, Barack 166, 167
obesity, and corruption 7–9
Oman, mosques 83
online shopping 143, 144
opioids, and coronavirus pandemic 13
oratory, Biden 166–7
Orthodox Christianity
American politicians 86
Russia 93
Osaka, cost of living 44
Oscars 183
osteichthyans 202
Osthagen, Andreas 68

P
Pagel, Mark 169
Pakistan
blasphemy laws 93
freedom 81
gender life-expectancy gap 112
pandemic *see* covid-19 pandemic

parents, Bank of Mum and Dad 23–5
Paris, cost of living 44
Patel, Dev 14–15
pay-transparency laws 110
Pegasus 234
penalties 213–14
Penrose, Roger 200, 201
PepsiCo 232–3
Le Père Noël est une ordure 240
Perfume Genius 230
period poverty 102–3
Peru, maritime boundaries 68
Petrobras 46, 47
Pfizer/BioNTech vaccine 124
Philadelphia 143
Philippines
  democracy 71
  maritime boundaries 68
  nurses 126–7
Pholensa, Dalaphone 39
placoderms 202
"Plants vs Zombies: Battle for Neighborville" 136–7
Poland
  Christmas television 240
  women's employment 107
Polisario Front 41, 42, 43
politicians
  corruption 7–9, 33–4, 46–8
  religious affiliation 85–7
Pompeo, Mike 180
Pop Smoke 227–8
population
  America 114–16
  predicting 99–101
Portugal, women's employment 107
Portuguese language 177
postage stamps 211–12, 242
posthumous albums 227–8
Powell, Jerome 150
predatory journals 204–5
Premier League 213–14
Presley, Elvis 97
Protestantism
  American politicians 86
  in China 90, 91
Przybylski, Andrew 137
Pulitzer prize 182–3
Puth, Charlie 230
Putin, Vladimir 7

Q
Qaboos, Sultan 83
Qatar, freedom 82
The Queen's Gambit 223

R
Raine, Elizabeth 198–9
Ramaphosa, Cyril 34
rappers 227–8
Rashford, Marcus 213
Rau, Raghavendra 154
recency bias 182–3
recycling 187–8
Red Bull cars 221
Reekie, Jemma 218
referees 209–10, 213
religion
  American politicians 85–7
  blasphemy laws 93–4
  and French *laïcité* 88–9
  see also Christianity; Islam
religious terrorism, America 65
Republic of China *see* Taiwan
RNA/DNA vaccines 124–5
road deaths 119–20

Roberts, Felicia 21–2
robots 161–3
Rogen, Seth 12
Rogers, Ginger 237
Rous, Peyton 200
Rousseff, Dilma 47
royal family 52–3
running shoes 218, 219
R.U.R. ("Rossum's Universal Robots") (Capek) 161–3
Russia
  army 50
  and Baltic states 37–8
  blasphemy laws 93
  Christmas television 240
  democracy 71
  economy 17
  education 15
  Gamaleya vaccine 124
  gender life-expectancy gap 111, 112
  maritime boundaries 69
  obesity and corruption 7, 8
  see also Soviet Union

S

"Sacred (Fiben Remix)" (Erasure) 231
Sahrawis 41–2
salary history bans (SHBs) 109–110
Samarkand 84
San Francisco, cost of living 45
Sandefur, Justin 14–15
Sanskrit 168
Sapir-Whorf hypothesis 174–5
SARS-CoV-2
  vaccines 123–5
  see also covid-19 pandemic
Saudi Arabia
  blasphemy laws 93
  freedom 81, 82
Saul, Ben 68
Schneider, Romy 239
Schuermann, Joshua 230–31
Schumacher, Michael 220, 221, 222
sea boundaries 67–9
sea ice 189–90
secularism 88–9
self-help books 35–6
sequoias 197
Seven Years as an Educated Youth (Xi Jinping) 36
Shakespeare, William 7
Shakur, Tupac 227
Shanghai 142
sharks 202–3
Shelley, Mary 161
Shenzhen 142
Shiga Naoya 160
shoes 218, 219
shopping, Black Friday 143–4
ShotLink 215, 216
Siberia
  forest fires 197
  heatwave 189
Sierra Leone, stamps 211, 212
Sinatra, Frank 98
Sinema, Kyrsten 85
Singapore, cost of living 44
Sinopharm vaccine 123
Sinovac vaccine 123
Siri 161
Sisi, Abdel-Fattah al- 79, 80, 83
Sissi 239
skeletons 202–3
Slovakia, women's employment 107

snow globes 236–8
Snow Patrol 229
The Snow Queen 240
Sommer, Marni 103
soup, ant-egg soup 39–40
South Africa
  economy 17
  prohibition 33–4
South America
  forests 196
  see also individual countries
South China Sea 68
South Korea
  democracy 71
  population predictions 100
  Twitter posts 165
  women's employment 107
sovereign grant 52
sovereignty, maritime
  boundaries 67–9
Soviet Union
  deposit refund scheme 187
  maritime boundaries 68
  see also Russia
Spain
  high-speed trains 54–5
  population predictions 100
  and Western Sahara 41, 42
  women's employment 107
Spanish language, Wikipedia 178
Spears, Britney 98
Spider-Man: Far from Home 234
Spratly archipelago 68
Stamperija 212
stamps 211–12
  Dinner for One 242
Standish, Jill 144
Steinberg, Philip 69
Stouraitis, Aris 154
strokes 133–5
Sudan, freedom 81, 82
Suleimani, Qassem 164, 165
Sun Tzu 36
super shoes 218, 219
Sussex, Duchess of 52, 53
Sussex, Duke of 52, 53
Sutton, Willie 19
Sweden
  Christmas television 239
  women's employment 107
Swedish language, Wikipedia 177
Swift, Taylor 237
swimming 219
Switzerland
  economy 17
  women's employment 107
Syria, freedom 81, 82

T

Taiwan
  fertility rates 100
  recognition 73–4
Tajikistan, obesity and
  corruption 7, 8
Tamerlane, Emperor 84
tampon tax 103
Tanaka Kane 133
Taoism 90
Tarlo, Emma 11
tea 33
Tel Aviv, cost of living 44
Telegram 76
television
  at Christmas 239–41
  most frequently aired
    programme 242–3
Temer, Michel 46–7
terrorism, America 64–6

Thailand, democracy 71
TIAA 107
Tianjin 141
Tigrayan People's Liberation Front (TPLF) 62–3
TikTok 229–30
Titilope, Olayinka 10, 11
Tomé, Carol 107
Tool 228
tourist visas 152–3
trade 145–6
trains, Spain 54–5
tramadol, and coronavirus pandemic 13
transport, high-speed trains 54–5
trees 196–7
Truelsen, Thomas 133
Trump, Donald
  The Art of the Deal 36
  and DC 60
  and FCPA 154
  immigration policy 115
  and religion 86
  and Russia 75
  speaking style 167
  and Western Sahara 41
Tsai Ing-wen 74
Tsugane Shoichiro 134
tuberculosis 111
Tullock paradox 154, 156
Turkey
  freedom 82
  women's employment 107
turkey meat 131–2
Turkmenistan, obesity and corruption 7, 8
Twitter 164–5

U
Uganda, covid-19 fatality rate 121, 122
Ukraine
  gender life-expectancy gap 111, 112
  obesity and corruption 7, 8
United Arab Emirates, freedom 82
United Nations (UN)
  and genocide 180–81
  population predictions 99–101
  Taiwan 73
  Western Sahara 41–2
United Nations Convention on the Law of the Sea (UNCLOS) 67, 68
United States (US)
  army 50
  and Baltic states 38
  Bank of Mum and Dad 23
  Black Friday 143–4
  and blasphemy laws 93, 94
  cannabis use 12
  Christmas television 239
  city cost of living 44, 45
  cola wars 232–3
  covid-19 fatality rate 121
  DC statehood 59–61
  democracy 71
  demographics 114–16
  digital currencies 149, 150
  economy 16–18
  education 15
  film production 234–5
  Foreign Corrupt Practices Act 154, 155
  gender life-expectancy gap 112
  gender pay gap 109–110

and genocide 180, 181
heatwaves 129, 130
maritime boundaries 68
military spending 49
period poverty 103
politicians' religious affiliation 85–7
Pulitzer prize 182–3
road deaths 119–20
snow globes 236–7
and Taiwan 73
terrorism 64–6
turkeys 131
video gaming 136–7
and Western Sahara 41, 43
women executives 107
women's employment 107, 108
UPS 107
Uyghurs 91, 180, 181
Uzbekistan, obesity and corruption 7, 8

V

Vacanze di Natale 239–40
vaccines 123–5
Vaiciunas, Zygimantas 38
Vaporflys 218, 219
Vatican, and Taiwan 74
Vegas weddings 97–8
La Vendetta (Balzac) 31
Venezuela, gender life-expectancy gap 112
Verkhoyansk 129
vertebrates 202
Veskimagi, Taavi 37
video calls 21–2
video games 136–7
Vietnam, gender life-expectancy gap 112
visas 152–3
Viva Las Vegas Wedding Chapel 97
Viviani, René 88

W

Walgreen Boots Alliance 107
The Wandering Earth 234
Wang Yi 91
war
Ethiopia 62–3
football and civil wars 5–6
Warden, May 242
Washington, DC 59–61
"We Didn't Start the Fire" (Billy Joel) 232
wealth
and education 14–15
royal family 52–3
weddings, in Las Vegas 97–8
weight, and corruption 7–9
Weiss-Wolf, Jennifer 103
West Germany
army 49
women's employment 104–6
see also Germany
Western Sahara 41–3
Wheeler Peak 195
White Christmas 239
Who Moved My Cheese? (Johnson) 35
wigs 10–11
Wikipedia 177–9
Williams cars 222
wine vocabulary 174–6
Winfrey, Oprah 52
"Without You" (Perfume Genius) 230
Wittgenstein, Ludwig 174, 176

Womé, Pierre 5
women
  employment in Germany 104–6
  fertility rates 99–100, 115
  gender pay gap 109–110
  glass-ceiling index 107–8
  life expectancy 111–13
  period poverty 102–3
  rights 81
  taking husbands' names 160
Wood, Paul 215
Woods, Tiger 215, 216
working mothers 104–6
  gender pay gap 109–110
World Bank
  and AfCFTA 145
  price comparison 16–18
World Health Assembly 73, 74
World Health Organisation (WHO)
  covid-19 pandemic 164, 165
  gaming disorder 137
  nurses 126, 127
  and Taiwan 73
World Values Survey (WVS) 70–72
Wrohlich, Katharina 105–6

X

Xi Jinping 36, 90
Xia Baolong 90
Xinjiang 91, 180, 181
XXXTentacion 227–8

Y

Yazidis 181
Yemen
  blasphemy laws 93
  freedom 81, 82
Yoshiro, Mori 108
"You'll Miss Me When I'm Not Around" (Grimes) 230
young people, and democracy 72
Yu Dan 36

Z

Zeno of Citium 26
Zhuravskaya, Ekaterina 75–6
Zoom meetings 21–2
Zuma, Jacob 34
Zurich, cost of living 44